Research Notes in Mathematics

Submission of proposals for consideration
Suggestions for publication, in the form of outlines and representative samples, are invited by the Editorial Board for assessment. Intending authors should approach one of the main editors or another member of the Editorial Board, citing the relevant AMS subject classifications. Alternatively, outlines may be sent directly to one of the publisher's offices. Refereeing is by members of the board and other mathematical authorities in the topic concerned, throughout the world.

Preparation of accepted manuscripts
On acceptance of a proposal, the publisher will supply full instructions for the preparation of manuscripts in a form suitable for direct photo-lithographic reproduction. Specially printed grid sheets are provided and a contribution is offered by the publisher towards the cost of typing. Word processor output, subject to the publisher's approval, is also acceptable.

Illustrations should be prepared by the authors, ready for direct reproduction without further improvement. The use of hand-drawn symbols should be avoided wherever possible, in order to maintain maximum clarity of the text.

The publisher will be pleased to give any guidance necessary during the preparation of a typescript, and will be happy to answer any queries.

Important note
In order to avoid later retyping, intending authors are strongly urged not to begin final preparation of a typescript before receiving the publisher's guidelines and special paper. In this way it is hoped to preserve the uniform appearance of the series.

Advanced Publishing Program
Pitman Publishing Inc
1020 Plain Street
Marshfield, MA 02050, USA
(tel (617) 837 1331)

Advanced Publishing Program
Pitman Publishing Limited
128 Long Acre
London WC2E 9AN, UK
(tel 01-379 7383)

Titles in this series

Nonlinear
variational
problems

A Marino, L Modica,
S Spagnolo & M Degiovanni (Editors)

University of Pisa

Nonlinear variational problems

Pitman Advanced Publishing Program
BOSTON · LONDON · MELBOURNE

PITMAN PUBLISHING INC
1020 Plain Street, Marshfield, Massachusetts 02050

PITMAN PUBLISHING LIMITED
128 Long Acre, London WC2E 9AN

Associated Companies
Pitman Publishing Pty Ltd, Melbourne
Pitman Publishing New Zealand Ltd, Wellington
Copp Clark Pitman, Toronto

First published 1985

AMS Subject Classifications: 35BXX, 35JXX, 35LXX

ISSN 0743-0337

Library of Congress Cataloging in Publication Data
Main entry under title:

Nonlinear variational problems.

    Bibliography: p.
    1. Differential equations, Partial—Addresses,
essays, lectures.   2. Variational inequalities
(Mathematics)—Addresses, essays, lectures.
I. Marino, A.
QA377.N68   1985   515.3'53   84-28978
ISBN 0-273-08670-7

British Library Cataloguing in Publication Data

Nonlinear variational problems.—(Research
  notes in mathematics, ISSN 0743-0337; 127)
  1. Variational inequalities (Mathematics)
  I. Marino, A.   II. Series
  515'.64   QA316

  ISBN 0-273-08670-7

Reproduced and printed by photolithography
in Great Britain by Biddles Ltd, Guildford

Contents

Preface

We have collected in this volume the invited talks and the short communications presented to the International Workshop on Nonlinear Variational Problems held at Isola d'Elba, Italy, in September 1983.

This workshop was, we believe, very useful in strengthening scientific bonds within the mathematical community, not least from the point of view of human relationships. The very high level of the scientific contributions was reinforced by the friendly atmosphere that prevailed throughout the workshop, which we remember with great pleasure and for which we are indebted to the speakers and to all the participants.

We also thank Marco Degiovanni, both for his help in organisational matters and for undertaking, with careful attention, the editorial work involved in preparing this volume for publication.

Pisa

March 1985

Antonio Marino

Luciano Modica

Sergio Spagnolo

INTERNATIONAL WORKSHOP ON NONLINEAR VARIATIONAL PROBLEMS
Hotel Lacona, Isola d'Elba, Italy, 26-30 September, 1983

SCIENTIFIC AND ORGANIZING COMMITTEE

Sergio Campanato
Ennio De Giorgi
Antonio Marino
Luciano Modica
Mauro Nacinovich
Giovanni Prodi
Sergio Spagnolo

INVITED SPEAKERS

Henri Berestycki (Paris) †
Haim Brezis (Paris)
Ronald J. DiPerna (Durham, N. Carolina)
Jens Frehse (Bonn) †
Sergiu Klainerman (New York)
Pierre-Louis Lions (Paris)
François Murat (Paris)
Louis Nirenberg (New York) †
Tatsuo Nishitani (Osaka)
Luc Tartar (Paris)

SHORT COMMUNICATIONS

A. Arosio
M. Biroli
G. Bottaro and P. Oppezzi
I. C. Dolcetta
M. Carriero and E. Pascali
E. Cavazzuti
R. de Arcangelis
M. Degiovanni

F. Donati
P. Donato
M. G. Garroni
M. Marino and A. Maugeri
A. Maugeri
S. Salerno
A. Valli
T. Zolezzi

† Text not submitted for publication in this volume.

F MURAT & L TARTAR
Optimality conditions and homogenization

In an optimization problem there are usually two important questions, often independent: existence of **a** solution and characterization of a solution.

The existence question usually involves some kind of compactness argument (even convexity arguments often rely on the fact that closed bounded convex sets in a reflexive Banach space are weakly compact) and the characterization involves computing derivatives or, more generally, computing the variations of some functionals. In some instances this computation is not easy and the theory of calculus of variations was certainly developed in order to understand this kind of question.

Where the variable is a domain, the computing of variations was done as early as 1905 by Hadamard, by pushing the boundary along the normal and then computing the induced variation of the functional. Turning this idea into theorems is not easy, and moreover already one can see another defect of this method: a given domain is compared only with domains of the same shape; it is not possible to make a hole inside the domain or to add a few small pieces far away by means of this technique. (Let us point out here that the classical methods dealing with stability analysis have the same defect; there is some hope of using the methods that we have developed to improve this kind of argument.)

The real difficulty lies in the fact that the set of domains, i. e. characteristic functions, does not possess natural paths from one domain to another: there is no manifold structure that enables us to use classical derivatives.

In the classical control problem, where the variable is a function of time, this difficulty has been overcome by Pontryaguin and his collaborators. It is now clear that they were first introducing generalized functions, as L. C. Young had done before, and were then computing classical derivatives.

In the problem where the variable appears as a domain and some partial differential equation is involved, there is another phenomenon that we discovered ten years ago (it was later called homogenization) : generalized domains appear which are the analogue of a mixture of two different materials and the effective properties of these mixtures have to be understood (they are not obtained by averaging certain quantities in more than one dimension).

If the method was already understood ten years ago (see Tartar [10], although we did not write up much of what was announced there), a lot remained to be understood about effective properties and it was only in 1981 that we found some of the characterization we had been looking for. We have started recently to write up most of the results accumulated during these years: we refer to Murat-Tartar [8] for a more complete exposition of the method and to Tartar [11] for the proof of the characterization.

The same kinds of idea were independently developed in the USSR by Lurie and his group [5], [6], [7]. These ideas have also been applied in a different context by R. Kohn and G. Strang [1], [2], [3], [4].

TWO MODEL PROBLEMS

We consider a bounded open domain Ω in $\mathbb{R}^N$ ($N = 2$ or 3 in applications). We want to choose a subset ω in Ω of given measure γ which will be optimal in some sense; no assumption of smoothness of ω is imposed (the phenomenon we want to examine here corresponds to having ω made of tiny pieces and this may be ruled out by some smoothness hypothesis).

Let us consider a physical problem of heat conduction: we fill ω with an isotropic material with heat conductivity α and the complement $\Omega \setminus \omega$ with an isotropic material with heat conductivity β. Assume that $0 < \alpha < \beta$ and let χ be the characteristic function of ω and denote a by

$$a(x) = \alpha \chi(x) + \beta(1 - \chi(x)).\tag{1}$$

We solve the equation

$$\begin{cases} -\text{div (a grad u)} = 1 \\ u\big|_{\partial\Omega} = 0 \ \ \text{i.e.} \ \ u \in H_0^1(\Omega) \end{cases}\tag{2}$$

2

which gives the equilibrium temperature $u(x)$ corresponding to a uniform heat source inside Ω, the boundary of Ω being held at temperature 0.

Consider the functional

$$J(\omega) = \int_\Omega u(x)\,dx \tag{3}$$

which, to a multiplicative factor, gives the average temperature in Ω.

We are interested in finding the extremal values of $J(\omega)$; we consider thus the two following problems:

Problem 1: Find ω which maximizes $J(\omega)$.

Problem 2: Find ω which minimizes $J(\omega)$.

Of course we impose on ω the constraint

$$\text{meas}(\omega) = \gamma \tag{4}$$

(without (4) the answer to Problem 1 is $\omega = \Omega$ and the answer to Problem 2 is $\omega = \phi$).

In general, no optimal solution exists. The intuitive reason for this is the following: in order to maintain the heat inside Ω in Problem 1, one should screen the effect of the good conductor. This is done by placing within some region a mixture of the two materials, made by alternating thin slices of each material perpendicularly to the direction of the heat flux; for Problem 2 one should enhance the effect of the good conductor and slices (more generally, fibres in dimension 3 or more) should be used in the direction of the flux.

SOME RESULTS IN HOMOGENIZATION

If we consider a sequence of characteristic functions χ_ε we can extract a weakly $\star$ converging subsequence

$$\chi_\varepsilon \to \theta \text{ in } L^\infty(\Omega) \text{ weak } \star \tag{5}$$

3

and the weak limit is characterized by

$$\begin{cases} 0 \le \theta(x) \le 1 \text{ a. e.} \\ \int_{\Omega} \theta(x)\, dx = \gamma . \end{cases} \qquad (6)$$

A subsequence of the corresponding sequence of solutions u_{ε} will converge weakly to u_0 but, except in one dimension, u_0 cannot be obtained from the knowledge of θ alone. We need to understand the effective heat conductivity of the limiting material, which is a mixture in regions where $0 < \theta(x) < 1$, and $\theta(x)$ is the local proportion of the material of conductivity α used in the mixture.

Mathematically this can be stated by the following, now classical

Proposition 1: If a_{ε} is a sequence of functions satisfying

$$\alpha \le a_{\varepsilon}(x) \le \beta \quad \text{a. e} \qquad (7)$$

then there exists a subsequence a_{η} and a (measurable in x) symmetric positive definite matrix $a_{\star}$ satisfying

$$\alpha I \le a_{\star}(x) \le \beta I \quad \text{a. e} \qquad (7')$$

such that, for every $f \in L^2(\Omega)$ (or more generally $f \in H^{-1}(\Omega)$), the solution u_{η} of

$$-\text{div}(a_{\eta} \text{ grad } u_{\eta}) = f \text{ and } u_{\eta} \in H^1_0(\Omega) \qquad (8)$$

converges weakly in $H^1_0(\Omega)$ to the solution u_0 of

$$-\text{div}(a_{\star} \text{ grad } u_0) = f \text{ and } u_0 \in H^1_0(\Omega). \qquad (8')$$

After we had rediscovered this result, which had been first obtained by S. Spagnolo [9], our first objective was to understand the relation between θ and $a_{\star}$ in the special situation where a_{ε} has the form (1) and θ is defined by (5); it took us a long time before we could characterize them by:

4

<u>Proposition 2</u>: Under hypotheses (1), (5), (6) the matrix $a_\star(x)$ described in Proposition 1 has (a.e.) its eigenvalues $\lambda_1, \ldots, \lambda_N$ satisfying

$$\mu_-(\theta(x)) \leq \lambda_j(x) \leq \mu_+(\theta(x)) \quad \text{a.e.} \quad \forall j \tag{9}$$

$$\begin{cases} \displaystyle\sum_j \frac{1}{\lambda_j - \alpha} \leq \frac{1}{\mu_-(\theta) - \alpha} + \frac{N-1}{\mu_+(\theta) - \alpha} \\[3mm] \displaystyle\sum_j \frac{1}{\beta - \lambda_j} \leq \frac{1}{\beta - \mu_-(\theta)} + \frac{N-1}{\beta - \mu_+(\theta)} \end{cases} \tag{10}$$

where (11) $\mu_-(\theta) = \left(\dfrac{\theta}{\alpha} + \dfrac{1-\theta}{\beta}\right)^{-1}$ and $\mu_+(\theta) = \theta \alpha + (1-\theta)\beta$. Conversely, if a matrix $a_\star$ has its eigenvalues satisfying (9), (10), then one can find a corresponding sequence χ_ε.

An important case is the one where, locally, $a_\varepsilon(x)$ depends only on one variable $(x \cdot e)$, where e is some unit vector; then $a_\star(x)$ has e as eigenvector corresponding to the eigenvalue $\mu_-(\theta(x))$ and the orthgonal of e as eigenspace with eigenvalue $\mu_+(\theta(x))$.

For many problems this remark, together with (9), gives the answer, and the more technical information (10) is not needed. This is the case here for each of our two problems.

RELAXED PROBLEMS

We have a relaxed problem consisting in replacing a domain ω, or its characteristic function χ, by a generalized domain $(\theta, a_\star)$ corresponding to a mixture. We then solve

$$-\text{div}(a_\star \,\text{grad}\, u) = 1 \quad \text{and} \quad u \in H_0^1(\Omega) \tag{12}$$

and compute the new functional

$$\tilde{J}(\theta, a_\star) = \int_\Omega u(x)\,dx. \tag{13}$$

The relaxed problem associated to Problem 1 consists in maximizing $\tilde{J}$ under the constraints (6), (9), (10), (and minimizing $\tilde{J}$ for Problem 2). This new problem has a solution, thanks to Propositions 1 and 2; moreover, computing variations of $\tilde{J}$ is not difficult: the inequalities (9), (10) actually define for each θ a convex set of matrices $a_\star$. A straightforward computation shows that

$$\delta\tilde{J} = -\int_\Omega (\delta a_\star \text{ grad u, grad u}) \, dx \tag{14}$$

(in the particular case of the functional $\tilde{J}$ that we consider here, an adjoint state, introduced to avoid explicit use of δu, coincides with u).

Because of (14) one sees that the optimal $a_\star(x)$ can be found in term of $\theta(x)$ (at points where grad u $\neq 0$) : $a_\star(x)$ will have grad u as an eigenvector with eigenvalue $\mu_-(\theta)$ in Problem 1 and $\mu_+(\theta)$ in Problem 2.

Because of the preceding remark we can then introduce a second relaxed problem, simpler than the first one, which only involves θ:

Problem 1': Given θ satisfying (6), solve the equation

$$-\text{div}(\mu_-(\theta)\,\text{grad u}) = 1 \quad \text{and} \quad u \in H_0^1(\Omega) \tag{15}$$

and maximize, under constraint (6), the functional given by

$$J_1(\theta) = \int_\Omega u(x)\,dx. \tag{16}$$

Problem 2': Given θ satisfying (6), solve the equation

$$-\text{div}(\mu_+(\theta)\,\text{grad v}) = 1 \quad \text{and} \quad v \in H_0^1(\Omega) \tag{17}$$

and minimize, under constraint (6), the functional given by

$$J_2(\theta) = \int_\Omega v(x)\,dx. \tag{18}$$

Problem 1' can be shown to be equivalent to maximizing a concave function of (u, θ) and the optimality condition, which is just a way of writing again (14), then turns out to be a necessary and sufficient condition for optimality; with the introduction of a Lagrange multiplier it becomes

$$\begin{cases} \exists\, C_1 \geq 0: & \theta(x) = 0 \Rightarrow \mu_-(\theta)\,\big|\text{grad } u\big| \leq C_1 \\ & 0 < \theta(x) < 1 \Rightarrow \mu_-(\theta)\,\big|\text{grad } u\big| = C_1 \\ & \theta(x) = 1 \Rightarrow \mu_-(\theta)\,\big|\text{grad } u\big| \geq C_1 \,. \end{cases} \tag{19}$$

Problem 2' can be treated by applying a min-max theorem to a convex-concave function and here also we obtain a necessary and sufficient condition for optimality:

$$\begin{cases} \exists\, C_1 \geq 0 & \theta(x) = 0 \Rightarrow \big|\text{grad } u\big| \geq C_1 \\ & 0 < \theta(x) < 1 \Rightarrow \big|\text{grad } u\big| = C_1 \\ & \theta(x) = 1 \Rightarrow \big|\text{grad } u\big| \leq C_1 \,. \end{cases} \tag{20}$$

Because of these optimality conditions, one can show that for most Ω an optimal solution has to take values between 0 and 1 and no classical solution ω, corresponding to θ being a characteristic function, exists.

Even when a classical solution exists, the optimality conditions obtained here are much stronger than what would have been obtained by just moving the boundary of ω. (19), (20) give necessary conditions on $\big|\text{grad } u\big|$ at each point of Ω; the method of moving the boundary of ω only gives conditions on $\partial\omega$.

It is worth noticing that application of numerical techniques to look for an optimal solution may lead to unstable results in the case where no classical solution exists; doing the complete analysis and introducing the problem with θ indicates new ways of approximating the solution which then appear to be more stable.

A similar remark holds for some inverse problems where, in some cases, no solution exists: introducing the right kind of homogenization may be a way to overcome this difficulty.

REFERENCES

1. R. Kohn, G. Strang: Structural design optimization, homogenization and relaxation of variational problems, in Macroscopic Properties of Disordered Media, Burridge, Childress, Papanicolaou ed., Lecture Notes in Physics 154, Springer (1982) 131-147.

2.   R. Kohn, G. Strang: Explicit relaxation of a variational problem in optimal design. Bull. A. M. S. 9 (1983) 211-214.

3.   R. Kohn, G. Strang: Optimal design for torsional rigidity, in Hybrid and Mixed Finite Element Methods, Atluri, Gallagher, Zienkiewicz ed., Wiley (1983) 281-288.

4.   R. Kohn, G. Strang: Optimal design and relaxation of variational problems (to appear).

5.   K. A. Lurie, A. V. Cherkaev: Optimal structural design and relaxed controls. Opt. Control. Appl. Math. 4 (1983) 387-392.

6.   K. A. Lurie, A. V. Cherkaev: Exact estimates of conductivity of composites formed by two isotropically conducting media taken in prescribed proportion, Proc. Roy. Soc. Edimburg A (to appear).

7.   K. A. Lurie, A. V. Cherkaev, A. V. Fedorov: Regularization of optimal design problems for bars and plates I, II, J. Opt. Th. Appl. 37 (1982) 499-522, 523-543.

8.   F. Murat, L. Tartar: Calcul des variations et homogénéisation, in Collection de la Direction des Etudes et Recherches d'Eléctricité de France, Eyrolles, Paris (1984), (Cours de l'Ecole d'Eté CEA. EDF. INRIA sur l'homogénéisation) (to appear).

9.   S. Spagnolo: Sulla convergenza di soluzioni di equazioni paraboliche ed ellittiche. Ann. Sc. Norm. Sup. Pisa 22 (1968) 577-597.

10.   L. Tartar: Problèmes de contrôle de coefficients dans des équations aux dérivées partielles, in Control Theory, Numerical Methods and Computer Systems Modelling. Lecture notes in Economics and Mathematical Systems 107, Springer (1974) 420-426.

11.   L. Tartar: Estimations fines de coefficients homogénéisés, in Ennio de Giorgi Colloquium, Krée ed., Research Notes in Mathematics 125, Pitman (1985, to appear).

François Murat   Luc Tartar
Analyse Numérique   C. E. A. Limeil and
Tour 55-65, 5e étage   Ecole Polytechnique
Université de Paris VI   France
4 Pl. Jussieu, 75230 Paris Cedex 05
France

T NISHITANI
The Cauchy problem for effectively hyperbolic operators

1. INTRODUCTION

Let P be the principal symbol of a hyperbolic differential operator. At a double characteristic point of P, its Taylor expansion begins with a quadratic form in the cotangent bundle. The coefficient matrix of the Hamiltonian system with this quadratic Hamiltonian is called the fundamental (or Hamilton) matrix. The eigenvalues of the fundamental matrix lie on the imaginary axis with the possible exception of one pair λ, $-\lambda$ with $\lambda > 0$. If the fundamental matrix has non-zero real eigenvalues, P is said to be an effectively hyperbolic operator ([3], [5]).

Ivrii and Petkov conjectured in [5], [6] that the Cauchy problem for effectively hyperbolic operators is C^∞ well posed for any lower order term; that is, an effectively hyperbolic operator is strongly hyperbolic.

Our purpose in this paper is to show that the conjecture is true. For another approach to this problem, we refer to Iwasaki [8], [9].

In Section 2, we show that the solvability of the Cauchy problem for hyperbolic pseudodifferential operators is reduced to the existence of parametrices of its factors with finite propagation speed of wave front sets. In Section 3, we give reduced forms of effectively hyperbolic pseudodifferential operators of second order by using canonical transformations in the cotangent bundle which do not depend on time and its dual variables. In Section 4, we derive microlocal energy estimates for such reduced effectively hyperbolic operators and we prove the existence of parametrices with finite propagation speed of wave front. In Section 5, we study in more detail the wave front sets of solutions of effectively hyperbolic equations.

2. MICROLOCAL PARAMETRIX WITH FINITE PROPAGATION SPEED OF WF

We consider the symbol

$$P(x, \xi) = \xi_0^m + \sum_{j=1}^{m} a_j(x, \xi') \xi_0^{m-j},$$

where $x = (x_0, x')$, $x' = (x_1, \ldots, x_d)$, $\xi = (\xi_0, \xi')$, $\xi' = (\xi_1, \ldots, \xi_d)$ and $a_j(x, \xi')$ are classical pseudodifferential symbols of degree j defined in a conic neighbourhood W of $\rho = (0, \bar{\xi}')$, depending smoothly on x_0. Extending $a_j(x, \xi')$ to outside W, we denote it also by $a_j(x, \xi')$. Hereafter, we say that the operator $P(x, D)$ with such a symbol is a monic x_0-differential operator with classical pseudodifferential coefficients of order m.

Let I be an open interval in $\mathbb{R}$ containing 0. We denote by $C_+^\ell(I, H^q)$ the set of all $v \in C^\ell(I, H^q)$ vanishing in $x_0 < 0$, where H^q stands for the usual Sobolev space $H^q(\mathbb{R}^d)$. For $v \in C^0(I, H^q)$, we write $||v||_q(x_0)$ instead of $||v(x_0, \cdot)||_q$ and sometimes we abbreviate it to $||v||_q$. We say that an operator G acting from $C^0(I, H^p)$ into $C^0(I, H^q)$ is a plus operator if Gv vanishes in $x_0 < 0$ for any $v \in C_+^0(I, H^p)$. We shall say that R is of Volterra type if, for any $\ell \in \mathbb{N}$, we have

$$D_0^\ell(Rv) = \sum_{j=0}^{\ell} r_{\ell,j} D_0^j v + \tilde{r}_\ell v,$$

where $r_{\ell,j}$ and $\tilde{r}_\ell$ satisfy the following conditions:

$r_{\ell,j}$ belongs to $S^{-\infty}$ and depends smoothly on $x_0 \in I$;   (2.1)

$\tilde{r}_\ell$ is a plus operator satisfying the following inequalities with a positive constant $\delta(R)$,

$$||\tilde{r}_\ell v||_p^2(t) \leq C_{p,q} \int_0^t ||v||_q^2 dx_0, \quad 0 \leq t \leq \delta(R), \qquad (2.2)$$

for any $p, q \in \mathbb{R}$, $v \in C_+^0(I, H^q)$.

Let P be a monic x_0-differential operator with classical pseudodifferential coefficients of order m. Let Γ be an open conic set in $T^*\mathbb{R}^d \backslash 0$. Assume that

G is a plus operator from $C^0(I, H^q)$ to $C^0(I, H^{q+n})$ for any $q \in \mathbb{R}$ (n denotes the loss of regularity of solutions, cf. (2.4) below). We shall say that G is a parametrix of P in Γ with finite propagation speed of WF if G satisfies the following conditions:

for any $g(x', \xi') \in S^0_{1,0}$ with cone supp$[g] \subset\subset \Gamma$, we have $PGg = g + R$

with Volterra type R; $\hspace{9cm}$ (2.3)

for any $q \in \mathbb{R}$, $v \in C^0_+(I, H^q)$, we have

$$\sum_{j=0}^{\ell} || D_0^j Gv ||_q^2 (t) \leq C_q \{ \sum_{j=0}^{\ell-m} || D_0^j v ||_{q+n}^2 (t) + \int_0^t || v ||_{q+n}^2 dx_0 \};\hspace{2cm} (2.4)$$

for every open conic set Γ_0, Γ_1 in $T^\star \mathbb{R}^d \setminus 0$ with $\Gamma_0 \subset\subset \Gamma_1 \subset\subset \Gamma$, there is a positive $\delta = \delta(\Gamma_0, \Gamma_1)$ such that, for any $g(x', \xi')$, $h(x', \xi') \in S^0_{1,0}$ with cone supp$[g] \subset \Gamma_0$, $h \equiv 0$ in Γ_1, we have

$$\sum_{j=0}^{m-1} || D_0^j hGgv ||_p^2 (t) \leq C_{p,q} \int_0^t || v ||_q^2 dx_0, \quad 0 \leq t \leq \delta(\Gamma_0, \Gamma_1),\hspace{1.5cm} (2.5)$$

for any $p, q \in \mathbb{R}$, $v \in C^0_+(I, H^q)$.

<u>Remark 2.1</u> It is clear that the existence of a parametrix of P in Γ does not depend on extensions of P to outside Γ.

<u>Remark 2.2</u> (2.5) shows that if WF(v) is contained in Γ_0, then WF(Gv) is contained in Γ_1 after the time t, $0 \leq t \leq \delta(\Gamma_0, \Gamma_1)$.

The proofs of the following propositions are standard. We refer, for example, to [1], [2].

<u>Proposition 2.1</u> Let P_i (i = 1, 2) be monic x_0-differential operators with classical pseudodifferential coefficients of order m_i. If P_i has a parametrix in Γ with finite propagation speed of WF, then, for any $\tilde{\Gamma} \subset\subset \Gamma$, $P_1 P_2$ has a parametrix in $\tilde{\Gamma}$ with finite propagation speed of WF.

Let $\chi; T^\star \mathbb{R}^d \to T^\star \mathbb{R}^d$ be a local canonical diffeomorphism depending smoothly on x_0 which is defined in an open neighbourhood of $\{0\} \times \Gamma_x$ in

$\mathbb{R} \times T^{\star} \mathbb{R}_x^d$. Let F be an elliptic Fourier integral operator associated with χ. We assume that

$$PFg = FP^{\chi}g + \sum_{j=0}^{m-1} r_j D_0^j$$

for any $g(x', \xi') \in S_{1,0}^0$ with cone $\text{supp}[g] \subset\subset \Gamma_y$, where P^{χ} is a monic x_0-differential operator with classical pseudodifferential coefficients of order m, and r_j belongs to $S^{-\infty}$ depending smoothly on x_0.

Proposition 2.2 Assume that P^{χ} has a parametrix in Γ_y with finite propagation speed of WF. Then, if Γ_y is sufficiently small, for any $\tilde{\Gamma}_x \subset\subset \chi^{-1}(\{0\} \times \Gamma_y)$, P has a parametrix in $\tilde{\Gamma}_x$ with finite propagation speed of WF.

Proposition 2.3 Let P be a monic x_0-differential operator with classical pseudodifferential coefficients of order m. Suppose that, for each $(0, \xi')$ $(|\xi'| = 1)$, there exists an open conic neighbourhood in which P has a parametrix with finite propagation speed of WF. Then the Cauchy problem for P is locally solvable in a neighbourhood of the origin with data on $x_0 = 0$.

3. REDUCED FORMS OF EFFECTIVELY HYPERBOLIC OPERATORS

Let $P(x, \xi)$ be a monic x_0-differential symbol with classical pseudodifferential coefficients of order m defined in a conic neighbourhood of $\rho = (0, \bar{\xi}')$. We say that P is effectively hyperbolic at ρ if the principal symbol P_m of P satisfies the following conditions:

$P_m(x, \xi)$ is hyperbolic with respect to dx_0, that is,
$P_m(x, \xi_0, \xi') = 0$ has only real roots in ξ_0 for any
(x, ξ') near ρ; (3.1)

if $dP_m(0, \xi_0, \bar{\xi}') = 0$, then the fundamental matrix
$F_{P_m}(0, \xi_0, \bar{\xi}')$ of P_m at $(0, \xi_0, \bar{\xi}')$ has non-zero real (3.2)
eigenvalues.

<u>Remark 3.1</u> If $P_m(0, \xi_0, \bar{\xi}') = 0$ has a root $\bar{\xi}_0$ of multiplicity greater than three, then the fundamental matrix of P_m at $(0, \bar{\xi})$ is the zero matrix ([3], [5]). Therefore, modulo an x_0-differential operator with $S^{-\infty}$ coefficients, microlocally, P is a product of effectively hyperbolic x_0-differential operators of order two or one.

Let $Q(x, \xi')$ be a symbol defined in a conic neighbourhood of $\rho = (0, \bar{\xi}')$ which is non-negative, homogeneous of degree 2. In the following, $\{,\}$ denotes the Poisson bracket.

<u>Lemma 3.1</u> Assume that $Q(\rho) = 0$, $\partial_0^2 Q(\rho) > 0$. Then there is a local homogeneous canonical diffeomorphism in $T^\star \mathbb{R}^d$ taking $\hat{\rho}$ to ρ under which $Q(x, \xi')$ takes one of the following forms in a conic neighbourhood of $\hat{\rho}$:

$$\sum_{i=0}^{p-1} q_i(x, \xi')(x_i - x_{i+1})^2 + \sum_{i=1}^{p} r_i(x, \xi')\xi_i^2$$

$$+ \{(x_p - \phi(x'', \xi''))^2 + \psi(x'', \xi'')\} q_p(x, \xi')$$

with $\{\phi, \{\phi, \psi\}\}(\hat{\rho}) = 0$, $0 \leq p \leq d-1$, $\qquad\qquad (3.3)_p$

$$\sum_{i=0}^{p-1} q_i(x, \xi')(x_i - x_{i+1})^2 + \sum_{i=1}^{p} r_i(x, \xi')\xi_i^2 + g(x_p, x'', \xi'') r_{p+1}(x, \xi'),$$

with $\{\xi_p, \{\xi_p, g\}\}(\hat{\rho}) = 0$, $1 \leq p \leq d-1$, $\qquad\qquad (3.4)_p$

where q_i, r_i are positive, homogeneous of degree 2, 0 respectively, ψ, g are non-negative, vanishing at $\hat{\rho}$, ϕ is homogeneous of degree 0 and $x'' = (x_{p+1}, \ldots, x_d)$, $\xi'' = (\xi_{p+1}, \ldots, \xi_d)$. Moreover, in $(3.3)_p$, it may be assumed that ϕ satisfies $d(\phi - x_{p+1})(\hat{\rho}) = 0$ or $d\phi(\hat{\rho}) = 0$.

The proof is by repeated use of the Malgrange preparation theorem and an existence theorem of homogeneous Darboux coordinates in $T^\star \mathbb{R}^d$ (for example, see Proposition 3.11 in [11]).

Now we consider the following effectively hyperbolic pseudodifferential symbol of second order,

$$P(x, \xi) = \xi_0^2 - Q(x, \xi').$$

Lemma 3.2 ([13]) Suppose that the fundamental matrix $F_p(0, 0, \bar{\xi}')$ has non-zero real eigenvalues. Then, in a conic neighbourhood of ρ, there exists a local homogeneous canonical diffeomorphism in $T^\star \mathbb{R}^d$ taking $\hat{\rho}$ to ρ under which $Q(x, \xi')$ is transformed to $(3.3)_p$ or to $(3.4)_p$ with $\Sigma_{i=1}^p r_i(\hat{\rho})^{-1} > 1$.

Remark 3.2 If $\partial_0^2 Q(\rho) > 0$ and the fundamental matrix $F_p(0, 0, \bar{\xi}')$ has no non-zero real eigenvalue, then it follows from Lemmas 3.1 and 3.2 that $Q(x, \xi')$ is transformed to $(3.4)_p$ with $\Sigma_{i=1}^p r_i(\hat{\rho})^{-1} \leqq 1$ by a homogeneous canonical transformation in $T^\star \mathbb{R}^d$.

Here we give some remarks on reduced forms of Lemma 3.2. Let $\Sigma = \{(x, \xi); P(x, \xi) = dP(x, \xi) = 0\}$ be the double characteristic set of $P(x, \xi)$ and denote $\text{Char} P = \{(x, \xi); P(x, \xi) = 0\}$.

First we introduce the hypersurfaces in $\mathbb{R} \times T^\star \mathbb{R}^d$:

$$S_\varepsilon ; x_0 - \phi(x'', \xi'') = \varepsilon \text{ in } (3.3)_p, \quad x_0 - R^{-1}\Sigma_{i=1}^p r_i^{-1} x_i = \varepsilon \text{ in } (3.4)_p$$

where $r_i = r_i(\hat{\rho})$, $R = \Sigma_{i=1}^p r_i^{-1}$ and ε is a positive parameter. Obviously S_0 contains Σ. Taking into account that $R > 1$, we can take $\kappa > 0$ so that

$$\kappa > 1 \text{ in } (3.3)_p, \quad R\kappa^2 < 1 \text{ and } \kappa < 1 \text{ in } (3.4)_p.$$

We define f by

$$f(x, \xi'; \delta) = -x_0 - \kappa(\mu|x'-y'|^2 + |x_{p+1} - y_{p+1}|^2 + \mu|\xi'||\xi'|^{-1} - \eta'|\eta'|^{-1}|^2)^{\frac{1}{2}},$$

$$f(x, \xi'; \delta) = -x_0 - \kappa(\Sigma_{i=1}^p r_i^{-1}|x_i - y_i|^2 + \mu|x'' - y''|^2 + \mu|\xi'||\xi'|^{-1} - \eta'|\eta'|^{-1}|^2)^{\frac{1}{2}},$$

in $(3.3)_p$, $(3.4)_p$ respectively, where $\delta = (y', \eta')$ and μ is a positive parameter.

Proposition 3.1 There are a conic neighbourhood W of $\hat{\rho}$ in $\mathbb{R} \times T^\star \mathbb{R}^d$ and a positive constant $\tilde{\kappa}$, $0 < \tilde{\kappa} < 1$ such that, for sufficiently small μ, we have in W that

$$T_{\hat{\rho}}S_0 \cap \{f(x, \xi'; \hat{\rho}) \geq 0\} = \{0\}, \quad \{f, Q\}^2 \leq 4\tilde{\kappa}Q .$$

Let $\gamma(s) = (x(s), \xi(s)) \subset \text{CharP}\backslash\Sigma$ be a bicharacteristic of P. If $f(x, \xi')$ satisfies the second inequality in Proposition 3.1 then it is easily verified that $\dot{x}_0(s)$ has a definite sign and

$$\frac{d}{ds}\phi(\gamma(s)) < 0 \text{ if } \dot{x}_0(s) > s, \quad \frac{d}{ds}\phi(\gamma(s)) > 0 \text{ if } \dot{x}_0(s) < 0,$$

where $\dot{x}_0(s)$ denotes differentiation with respect to s. Hence it follows that the projections on $\mathbb{R} \times T^\star\mathbb{R}^d$ of all bicharacteristics of P having the limit point $\hat{\rho}$ are contained in the cone $\{f(x, \xi') \geq 0\}$ which is transversal to S_0.

We shall use this hypersurface to obtain microlocal energy estimates for P in the next section.

4. MICROLOCAL ENERGY ESTIMATES

In this section, we derive microlocal energy estimates for reduced effectively hyperbolic operators of second order. For the detailed proofs, we refer to $[14]$. We also show that such operators have parametrices with finite propagation speed of WF. We consider the symbol

$$P(x, \xi) = \xi_0^2 - Q(x, \xi') + R(x, \xi')\xi_0, \tag{4.1}$$

where $Q(x, \xi')$, $R(x, \xi')$ are classical pseudodifferential symbols defined in a conic neighbourhood of $\hat{\rho} = (0, \hat{\xi}')$. We denote by $P_2(x, \xi)$, $Q_2(x, \xi')$ the principal symbols of P, Q respectively, and $P^s(x, \xi)$ stands for the subprincipal symbol of P. We assume that $Q_2(x, \xi')$ takes one of the reduced forms of Lemma 3.2.

To state the energy estimates, we introduce some notation. First, we make a change of scales of variables:

$$y = \mu^\sigma x = (\mu^{\sigma_0}x_0, \ldots, \mu^{\sigma_d}x_d), \quad 0 < \mu \leq 1.$$

We take σ so that $\sigma_j = \frac{1}{2}$, $0 \le j \le p$, $\sigma_j = 0$, $p+1 \le j \le d$ in the case of $(3.3)_p$ with $d\phi(\hat{\rho}) = 0$ and $\sigma_j = 1$, $0 \le j \le p+1$, $\sigma_j = \frac{1}{2}$, $p+2 \le j \le d$ if Q_2 has the form $(3.3)_p$ with $d\phi(\hat{\rho}) \ne 0$. In the case $(3.4)_p$, we take σ so that $\sigma_j = 1$, $0 \le j \le d$. Then we have

$$P_{(\mu)} = \mu^{2\sigma_0} P(\mu^{\sigma} x, \mu^{-\sigma} \xi) = \xi_0^2 - Q_2(x, \xi', \mu) + T_0(x, \xi', \mu) \xi_0$$
$$+ T_1(x, \xi', \mu),$$

where $T_i(x, \xi', \mu)$ is of degree i. Note that $Q_2(x, \xi', \mu)$ takes the following form in the case of $(3.3)_p$:

$$Q_2(x, \xi', \mu) = \mu^2 \sum_{i=0}^{p-1} q_i(x, \xi', \mu)(x_i - x_{i+1})^2 + \sum_{i=1}^{p} r_i(x, \xi', \mu) \xi_i^2$$

$$+ \{\mu^2(x_p - \phi(x'', \xi'', \mu))^2 + \psi(x'', \xi'', \mu)\} q_p(x, \xi', \mu),$$

with $q_i(x, \xi', \mu) = q_i(\mu^{\sigma} x, \mu^{-\sigma} \xi')$ if $d\phi(\hat{\rho}) = 0$ and $q_i(x, \xi', \mu) = \mu^2 q_i(\mu^{\sigma} x, \mu^{-\sigma} \xi')$ if $d\phi(\hat{\rho}) \ne 0$. In the case of $(3.4)_p$, we have

$$Q_2(x, \xi', \mu) = \mu^2 \sum_{i=0}^{p-1} q_i(x, \xi', \mu)(x_i - x_{i+1})^2 + \sum_{i=1}^{p} r_i(x, \xi', \mu) \xi_i^2$$

$$+ g(x_p, x'', \xi'', \mu) r_{p+1}(x, \xi', \mu),$$

with $r_i(x, \xi', \mu) = r_i(\mu x, \xi')$, $q_i(x, \xi', \mu) = q_i(\mu x, \xi')$.

Each symbol in the above expressions of $Q_2(x, \xi', \mu)$ is defined in some conic neighbourhood W_μ (depending on μ) of $\hat{\rho}$. We extend these symbols suitably to outside W_μ, and denote those also by the same letters.

Take $\chi_0(s) \in C^\infty(\mathbb{R})$ such that $\chi_0(s) = 1$ for $s \ge -1/4$, $\chi_0(s) = 0$ for $s \le -\frac{1}{2}$ and introduce the weight functions

$$J_{\pm}(x, \xi', \mu) = \pm \{2\chi_0(\pm Y(x, \xi', \mu)\langle \mu\xi' \rangle^{\frac{1}{2}}) - 1\} Y(x, \xi', \mu) + \langle \mu\xi' \rangle^{-\frac{1}{2}},$$

$$\langle \mu\xi' \rangle^2 = 1 + \sum_{j=1}^{d} \xi_j^2,$$

where

$$Y(x, \xi', \mu) = x_0 - \phi(x'', \xi', \mu) \quad \text{in } (3.3)_p,$$

$$Y(x, \xi', \mu) = x_0 - R^{-1} \sum_{i=1}^{p} r_i^{-1} x_i \quad \text{in } (3.4)_p,$$

(cf. Section 3). Next, choose $\chi(s) \in C^{\infty}(\mathbb{R})$ with $\chi(s) + \chi(-s) = 1$, $\chi(s) = 0$ for $s \leq -1$, $\chi(s) = 1$ for $s \geq 1$ and set

$$\alpha_n^{\pm}(x, \xi', \mu) = \chi(\pm n^{\frac{1}{2}} Y(x, \xi', \mu) \langle \mu \xi \rangle^{\frac{1}{2}}).$$

Using this notation, we introduce the following semi-norm,

$$\|u\|_{n+k, r}^2(x_0) = \|K_k^{-}(n-r) \alpha_n^{-} u\|^2(x_0) + \|K_k^{+}(n-r) \alpha_n^{+} u\|^2(x_0),$$

with $K_k^{-}(n-r) = op(\langle \mu \xi \rangle^{n+k} J_{-}(x, \xi', \mu)^{n-r})$, $K_k^{+}(n-r) = op(\langle \mu \xi \rangle^{k} J_{+}(x, \xi', \mu)^{n-r})$, where $op(a(x, \xi', \mu))$ denotes the pseudodifferential operator with symbol $a(x, \xi', \mu)$ and $\|\cdot\|$ is the norm in $L^2(\mathbb{R}^d)$.

<u>Remark 4.1</u> We note that the symbol

$$K_k^{-}(n)(x, \xi', \mu) \alpha_n^{-}(x, \xi', \mu) + K_k^{+}(n)(x, \xi', \mu) \alpha_n^{+}(x, \xi', \mu)$$

is equivalent to $\langle \mu \xi \rangle^{k+n}$ when $Y(x, \xi', \mu) \leq -c$ to $\langle \mu \xi \rangle^{k+(n/2)}$ when $|Y(x, \xi', \mu)| \leq c \langle \mu \xi \rangle^{\frac{1}{2}}$ and to $\langle \mu \xi \rangle^{k}$ when $Y(x, \xi', \mu) \geq c$ with arbitrary positive c.

Taking into account the identity $e^{-x_0 \theta} D_0 = (D_0 - i\theta) e^{-x_0 \theta}$, we derive energy estimates for $P_{(\mu), \theta}$ in place of $P_{(\mu)}$,

$$P_{(\mu), \theta} = (D_0 - i\theta)^2 - Q_2(x, D', \mu) + T_0(x, D', \mu)(D_0 - i\theta) + T_1(x, D', \mu).$$

Set

$$E_{n, s}(u, x_0) = \|u\|_{n+s+1, 0}^2 + \theta \|u\|_{n+s+1, -\frac{1}{2}}^2 + \theta^2 \|u\|_{n+s, 1}^2$$

$$+ \|(D_0 - i\theta) u\|_{n+s, 1}^2 + \theta \|(D_0 - i\theta) u\|_{n+2, \frac{1}{2}}^2,$$

$$e_{n, s}(u, x_0) = \|u\|_{n+s+1, -\frac{1}{2}}^2 + \theta \|u\|_{n+s, 1}^2 + \|(D_0 - i\theta) u\|_{n+s, \frac{1}{2}}^2$$

17

$$+ \sum {}^{\pm} \text{Re}(Q_2 \, K_0^{\pm}(n - \tfrac{1}{2}) \, \alpha_n^{\pm} \langle \mu D \rangle^s u, \, K_0^{\pm}(n - \tfrac{1}{2}) \, \alpha_n^{\pm} \langle \mu D \rangle^s u) \,,$$

$$|u|_{-L}^2(x_0) = \theta^{3/2} \|u\|_{-L}^2(x_0) + \|(D_0 - i\theta) u\|_{-L}^2(x_0) \,.$$

Then we have:

Theorem 4.1 ([14]) For every $n \geq C_0 C_1$, there is $\mu > 0$ such that, for any $L \geq 1 - s$, $s \in \mathbb{R}$ and $\theta \geq \theta_0(s, L)$, we have

$$c(s) \int_\tau^t \left\| P_{(\mu), \theta} u \right\|_{n+s, 0}^2 dx_0$$

$$+ c(n, s, L) \left\{ \int_\tau^t \left\| P_{(\mu), \theta} u \right\|_{-L}^2 dx_0 + \theta \, |u|_{-L}^2(\tau) \right\} + e_{n, s}(u, \tau)$$

$$\geq c \left\{ e_{n, s}(u, t) + \int_\tau^t E_{n, s}(u, x_0) \, dx_0 \right\} + \theta \int_\tau^t |u|_{-L}^2 dx_0 + |u|_{-L}^2(t) \,,$$

with $c > 0$, where $C_1 = |P^s(0, 0, \hat{\xi}')| + 1$, $C_0 = C_0(r_i(\hat{\rho}), q_i(\hat{\rho}))$.

Corollary 4.1 For any $s \in \mathbb{R}$, $u \in C_0^\infty(I \times \mathbb{R}^d)$, we have

$$c(n, s) \int e^{-2\theta x_0} \left\| P_{(1)} u \right\|_{n+s+1}^2 dx_0 \geq \theta^{3/2} \int e^{-2\theta x_0} \left\| D_0 u \right\|_s^2 dx_0 + \theta^3 \int e^{-2\theta x_0} \|u\|_s^2 dx_0,$$

where $n \geq C_0 C_1$, $\theta \geq \theta_0(n, s)$.

Remark 4.2 Corollary 4.1 shows that the loss of regularity of solutions is at most n.

Now we fix n, $\mu > 0$ so that the energy estimate in Theorem 4.1 holds and write $P = P_{(\mu)}$, $P_\theta = P_{(\mu), \theta}$, dropping μ. From Theorem 4.1, there exists an operator G from $C^0(I, H^q)$ into $C^0(I, H^{q+n})$ such that

$$PGv = v \quad \text{and} \quad Gv = 0 \quad \text{in } x_0 < t \quad \text{if } v = 0 \quad \text{in } x_0 < t.$$

To show that G is a parametrix with finite propagation speed of WF, we estimate the wave front sets of Gv in terms of Sobolev norms and derive the property (2.5). Let $\phi(x, \xi')$ be in $S_{1,0}^0$ depending smoothly on x_0. Following Ivrii [6], we define Φ by

$$\Phi = \begin{cases} \exp(-1/\phi) & \text{if } \phi > 0 \\ 0 & \text{if } \phi \leq 0. \end{cases} \tag{4.2}$$

It is clear that Φ, depending smoothly on x_0, belongs to $S^0_{1,0}$. Assume that

$$\partial_0 \phi = -1, \quad \{Q_2, \phi\}^2 \leq 4\kappa Q_2(x, \xi', \mu), \tag{4.3}$$

with a positive constant κ, $0 < \kappa < 1$. Then applying Theorem 4.1 and using the estimates of the commutator $[P_\theta, \Phi]$ (cf. Lemma 3.6 in $[7]$), we can show that the following inequality is valid for any $L \geq 1-\hat{s}$, s, $\hat{s} \in \mathbb{R}$, $s \geq \hat{s}$ that

$$\begin{aligned}
c(s) \int_\tau^t \left\|\!\left\|\!\left\| \Phi P_\theta u \right\|\!\right\|\!\right\|^2_{n+s,0} dx_0 &+ c(s,L) \int_\tau^t \left\| \Phi P_\theta u \right\|^2_{-L} dx_0 \\
&+ c(s) \{ e_{n,s}(\Phi u, \tau) + e_{n,s-1/4}(u,\tau) \} \\
&+ c(s) \{ e_{n,s-1/4}(u,t) + \int_\tau^t E_{n,s-1/4}(u,x_0) dx_0 \} \\
&+ c(\hat{s},L) \{ \int_\tau^t E_{n,\hat{s}}(u,x_0) dx_0 + e_{n,\hat{s}}(u,t) + \theta |u|^2_{-L}(\tau) \} \\
&\geq e_{n,s}(\Phi u, t) + \int_\tau^t E_{n,s}(\Phi u, x_0) dx_0 + \theta \int_\tau^t |u|^2_{-L} dx_0 + |u|^2_{-L}(t).
\end{aligned} \tag{4.4}$$

Without loss of generality, we may assume that the last term in the left-hand side of (4.4) is finite with some $\hat{s}$, L. Then one can improve the regularity of the cut-off solution Φu with $1/4$ in the semi-norm $\|\!\|\!\|\cdot\|\!\|\!\|$.

Successive application of this inequality gives

Lemma 4.1 Assume that ϕ satisfies (4.3) and $u \in C^1([T_-, T_+], H^q)$ with some $q \in \mathbb{R}$. Then if

$$\Phi P u \in C^0([T_-, T_+], H^s), \quad \Phi(T_-)u(T_-) \in H^{s+1}, \quad \Phi(T_-)D_0 u(T_-) \in H^s,$$

it follows that

$$\Phi_\nu u \in C^1([T_-, T_+], H^{s-n}),$$

for any $\nu > 0$, where Φ_ν is defined by (4.2) with $\phi_\nu = \phi + \nu$.

Take any two open conic sets Γ_0, Γ_1 in $T^\star \mathbb{R}^d$ with $\Gamma_0 \subset\subset \Gamma_1$ and two symbols $g(x', \xi')$, $h(x', \xi')$ belonging to $S_{1,0}^0$ such that the cone support of g is contained in Γ_0 and h vanishes in Γ_1. We choose the special ϕ in the inequality (4.4),

$$\phi = -x_0 + T + \varepsilon\left(\left|x'-y'\right|^2 + \left|\xi'\right|\left|\xi'\right|^{-1} - \eta'\left|\eta'\right|^{-1}\right|^2 + \delta\right)^{\frac{1}{2}}, \quad \delta > 0,$$

which satisfies (4.3) for sufficiently small $\varepsilon > 0$. Then, from (4.4), it follows that

Proposition 4.1   Let G, g, h be as above. Then there is a positive $\delta = \delta(\Gamma_0, \Gamma_1)$ such that

$$\sum_{j=0}^{1} \left\{ \left\|D_0^j h G g v\right\|_p^2(t) + \int_\tau^t \left\|D_0^j h G g v\right\|_p^2 dx_0 \right\} \leq C_{p,q} \int_\tau^t \left\|v\right\|_q^2 dx_0, \quad 0 \leq t - \tau \leq \delta,$$

for any $p \in \mathbb{R}$ and $v \in C^0([T_-, T_+], H^q)$ vanishing in $x_0 < \tau$.

This proposition shows that G is a parametrix of P in $\Gamma = W_\mu$ with finite propagation speed of WF. Hence the discussions in Sections 2 and 3 give

Theorem 4.2   Let P be a monic x_0-differential operator with classical pseudodifferential coefficients of order m defined in a conic neighbourhood of $\rho = (0, \bar{\xi}')$. Suppose that P is effectively hyperbolic at ρ. Then, in a conic neighbourhood of ρ, there exists a parametrix with finite propagation speed of WF.

The following theorem shows that the conjecture of Ivrii and Petkov is true.

Theorem 4.3 ([15])   Let P be a monic x_0-differential operator with classical pseudodifferential coefficients of order m. Assume that P is effectively hyperbolic at every $(0, \xi')$, $\left|\xi'\right| = 1$. Then the Cauchy problem for P is locally solvable in the class C^∞ in a neighbourhood of the origin with data on $x_0 = 0$.

5. WAVE FRONT SETS OF SOLUTIONS

In this section, we give some results concerning wave front sets of solutions of effectively hyperbolic operators of second order. The argument and the results throughout the sequel are microlocal.

We study the operator P in (4.1). The next theorem follows easily from Lemma 4.1.

Theorem 5.1 ($[16]$) Let ϕ be as in Lemma 4.1. Then it follows from

$$WF(Pu) \cap CharP \cap \{x_0 \geq T, \ \phi > 0\} = \emptyset, \ WF(u) \cap CharP \cap \{x_0 = T, \phi > 0\} = \emptyset,$$

that

$$WF(u) \cap CharP \cap \{x_0 \geq T, \ \phi > 0\} = \emptyset,$$

where $u \in \mathfrak{D}'$ and $CharP = \{(x, \ \xi) ; P_2(x, \ \xi) = 0\}$.

Remark 5.1 This theorem generalizes Theorem 3.8 in Ivrii $[7]$.

Fix $\varepsilon_0 < 0$ with sufficiently small modulus. We denote by $F^{\pm}_{\varepsilon}$ a mapping which assigns $(x, \ \xi) \in S_{\varepsilon_0}$ (which is defined in Section 3) to $(y, \ \eta) \in S_{\varepsilon}$ where $(x, \ \xi)$ and $(y, \ \eta)$ lie on the same null bicharacteristic of $\xi_0 \pm Q_2(x, \ \xi')^{\frac{1}{2}}$. Here we remark that Proposition 3.1 is valid uniformly with respect to ε, $\delta = (y', \ \eta')$ where ε is sufficiently small and δ is sufficiently near to $\hat{\rho}$. From this remark it follows that $F^{\pm}_{\varepsilon}$ defines a local diffeomorphism if $\varepsilon < 0$. Also it is not difficult to see that

$$\left| F^{\pm}_{\varepsilon_1} (x, \ \xi) - F^{\pm}_{\varepsilon_2} (x, \ \xi) \right| \leq C \left| \varepsilon_1 - \varepsilon_2 \right|, \ \varepsilon_0 \leq \varepsilon_i < 0,$$

with some constant C independent of $(x, \ \xi)$.

By $\Gamma^+(\rho)$ (resp. $\Gamma^-(\rho)$), we denote the union of all bicharacteristics $\gamma(s) \subset CharP\backslash\Sigma$, which have the limit point $\rho \in \Sigma$ when x_0 increases (resp. when x_0 decreases). Now we apply Theorem 5.1 with the special $\phi = f$ (f is the same as in Section 3) to the study of wave front sets (it may be necessary to modify f to be smooth). Then from the above discussions on $F_\epsilon^\pm$ and the arguments for f in Section 3, we have the following result (see also Melrose [12]).

Theorem 5.2 ([16]) Let $\rho \in \Sigma$, $u^\pm \in \mathcal{D}'$. Suppose that $\Gamma^\pm(\rho) \cap WF(u^\pm) = \emptyset$, $\rho \notin WF(Pu^\pm)$, then we have $\rho \notin WF(u^\pm)$.

Here we give some sufficient condition which assures the existence of two and only two distinct bicharacteristics that have limit points in the double characteristic set. We consider the symbol $P(x, \xi) = \xi_0^2 - Q(x, \xi')$ where $Q(x, \xi')$ is defined in a conic neighbourhood of ρ, non-negative and homogeneous of degree 2 in ξ'. We assume that $Q(\rho) = 0$ and P is effectively hyperbolic at ρ. Set

$$\mathrm{Tr}^+ F_p(\rho) = \Sigma \left| \text{pure imaginary eigenvalues of } F_p(\rho) \right|,$$

(the eigenvalues are repeated according to their multiplicities), then we have the following result.

Theorem 5.3 ([16]) Let $\pm\lambda$ $(\lambda > 0)$ be non-zero real eigenvalues of $F_p(\rho)$. Then, if $\mathrm{Tr}^+ F_p(\rho)/\lambda < 2$, there are two and only two distinct bicharacteristics of P which have the limit point ρ when x_0 increases. The same assertion is also true when x_0 decreases.

REFERENCES

1. J. J. Duistermaat, Fourier Integral Operators, Lecture Notes, Courant Institute of Mathematical Sciences, New York (1973).

2. L. Hörmander, Fourier integral operators I, Acta Math. 127 (1971), 79-183.

3. L. Hörmander, The Cauchy problem for differential equations with double characteristics, Jour. d'Analyse Math. 32 (1977), 118-196.

4. M. W. Hirsch and S. Smale, Differential Equations, Dynamical Systems, and Linear Algebra, Academic Press, New York (1974).

5.   V. Ja. Ivrii and V. M. Petkov, Necessary conditions for the Cauchy problem for non-strictly hyperbolic equations to be well posed, Russian Math. Surveys, 29 (1974), 1-70.

6.   V. Ja. Ivrii, Sufficient conditions for regular and completely regular hyperbolicity, Trans. Moscow Math. Soc., 1 (1978), 1-65.

7.   V. Ja. Ivrii, Wave fronts of solutions of certain pseudodifferential equations, Trans. Moscow Math. Soc., 39 (1981), 49-86.

8.   N. Iwasaki, Cauchy problem for effectively hyperbolic equations (a standard type), preprint (1983).

9.   N. Iwasaki, Cauchy problem for effectively hyperbolic equations (general case), preprint (1984).

10. H. Kumanogo, Pseudo Differential Operators, M. I. T. Press (1982).

11. R. B. Melrose, Equivalence of glancing hypersurfaces, Inventiones Math. 37 (1976), 165-191.

12. R. B. Melrose, The Cauchy problem for effectively hyperbolic operators, preprint (1983).

13. T. Nishitani, A note on reduced forms of effectively hyperbolic operators and energy integrals, Osaka Jour. Math. 21 (1984) (to appear).

14. T. Nishitani, Local energy integrals for effectively hyperbolic operators, I, II, Jour. Math. Kyoto Univ. 24 (1984) (to appear).

15. T. Nishitani, On the finite propagation speed of wave front sets for effectively hyperbolic operators, College Gen. Ed. Osaka Univ. 32 (1) (1983), 1-7.

16. T. Nishitani, On wave front sets of solutions for effectively hyperbolic operators, College Gen. Ed. Osaka Univ. 32 (2) (1983), 1-7.

Tatsuo Nishitani
Department of Mathematics
College of General Education
Osaka University
Toyonaka, Osaka
560 Japan

F MURAT
The Neumann sieve

In this paper, I report on joint work with Hedy Attouch, Alain Damlamian and Colette Picard. More details of our work and some extensions can be found in [1], [2], [7], [11], etc.

The model problem we shall study here is the following. Let Ω be an open bounded set of $\mathbb{R}^N$ ($N \geq 2$) with smooth boundary $\partial\Omega$. We suppose that Ω is divided into two parts by the hyperplane $\Sigma = \{x \mid x_N = 0\}$: the upper part,

$$\Omega_a = \Omega \cap \{x \mid x_N > 0\},$$

and the lower part,

$$\Omega_b = \Omega \cap \{x \mid x_N < 0\}.$$

(Throughout this paper we denote by the subscripts a and b what concerns the upper and lower parts, respectively.)

In addition, we assume the boundary $\partial\Omega$ to be transversal to Σ, so that Ω_a and Ω_b are smooth open bounded sets with some edges. Their boundaries $\partial\Omega_a$ and $\partial\Omega_b$ are divided into two parts, Γ and the "outer boundaries" $(\partial\Omega)_a$ and $(\partial\Omega)_b$:

$$
\begin{cases}
\Gamma = \overline{\Omega} \cap \Sigma = \{x \mid x \in \overline{\Omega}, \ x_N = 0\}, \\
(\partial\Omega)_a = \partial\Omega \cap \{x \mid x_N > 0\}, \\
(\partial\Omega)_b = \partial\Omega \cap \{x \mid x_N < 0\}, \\
\partial\Omega_a = \Gamma \cup (\partial\Omega)_a, \\
\partial\Omega_b = \Gamma \cup (\partial\Omega)_b.
\end{cases}
\tag{1}
$$

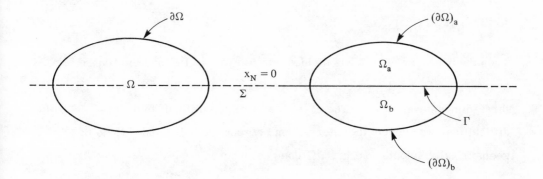

Figure 1 – The set Ω and its upper and lower parts

Let us now describe the sieve: it is a plane metal plate located on Σ and perforated with many small holes. For sake of simplicity, we assume here that the plate has zero thickness and the holes are regularly distributed on the metal and have the same shape $\mathcal{K}$ and the same size r^ε. Here ε is the parameter characterizing the distance between two adjacent holes. In mathematical terms, this means that the sieve represents the closed set Σ^ε of $\mathbb{R}^N$ defined by

$$\Sigma^\varepsilon = \{x \in \mathbb{R}^N | x_N = 0, \; x' - c \notin r^\varepsilon \mathcal{K} \qquad \forall c \in C^\varepsilon\} \tag{2}$$

Here $x' = (x_1, \ldots, x_{N-1})$ is the Euclidean coordinate in $\mathbb{R}^{N-1}$ and C^ε is the set of the centres of the holes:

$$C^\varepsilon = \{x \in \mathbb{R}^N | x_N = 0, \; x' = (2\varepsilon k_1, \ldots, 2\varepsilon k_{N-1}) \text{ with } k_j \in \mathbb{Z}\}. \tag{3}$$

Let us note that the distance between two adjacent centres is in fact 2ε. We denote by r^ε the radius of the hole (of course $r^\varepsilon < \varepsilon$) and by $\mathcal{K}$ the basic shape which is used to cut the small holes $c + r^\varepsilon \mathcal{K}$: $\mathcal{K}$ is an open connected set of $\mathbb{R}^{N-1}$, included in the ball $|x'| < 1$ and with smooth boundary. Let us finally define:

$$\begin{cases} \Gamma^\varepsilon = \bar{\Omega} \cap \Sigma^\varepsilon, \\ \Omega^\varepsilon = \Omega \setminus \Gamma^\varepsilon. \end{cases} \tag{4}$$

Then the boundary of $\partial\Omega^\varepsilon$ is divided into two parts, Γ^ε and the "outer boundary" $\partial\Omega$: $\partial\Omega^\varepsilon = \Gamma^\varepsilon \cup \partial\Omega$. (See Figures 2 and 3 for the two- and three-dimensional cases. The case $N \geq 4$ is too difficult to draw.) To avoid difficulties, we assume the intersection between the holes and $\partial\Omega$ to be geometrically smooth (e.g., $\partial\Omega$ does not meet $C^\varepsilon + r^\varepsilon \bar{\mathcal{H}}$).

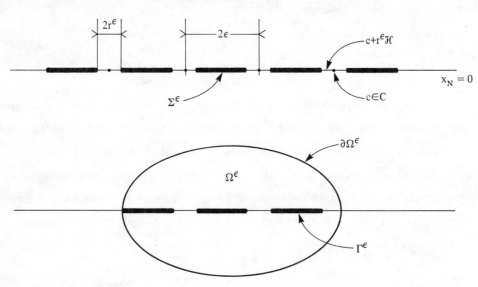

Figure 2 - The sieve Σ^ε and the open set Ω^ε in the two-dimensional case

We consider here the following boundary value problem:

$$\begin{cases} -\Delta u^\varepsilon + u^\varepsilon = f \text{ in } \Omega^\varepsilon, & (5) \\[2mm] \dfrac{\partial u^\varepsilon}{\partial n} = 0 \text{ on } \Gamma^\varepsilon, & (6) \\[2mm] u^\varepsilon = 0 \text{ on } \partial\Omega, & (7) \end{cases}$$

where f is a given function, $f \in L^2(\Omega)$. Our objective is to describe what happens as ε tends to zero. Of course the result is different according to the

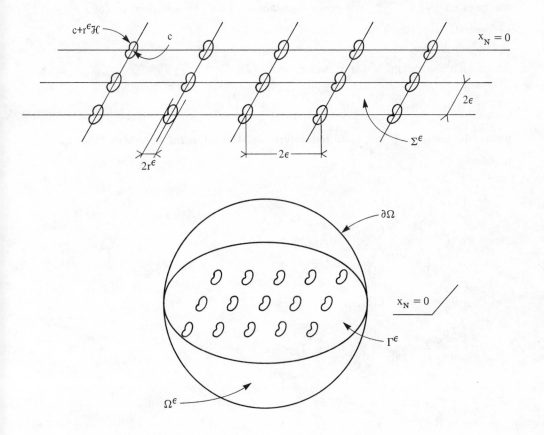

Figure 3 - The sieve Σ^ε and the open set Ω^ε in the three-dimensional case

behaviour of r^ε. The main result we prove is the following:

Theorem   Let us assume that r^ε is equivalent (in the sense $r^\varepsilon / r_\star^\varepsilon \to 1$ as $\varepsilon \to 0$) to $r_\star^\varepsilon$ defined by

$$\begin{cases} r_\star^\varepsilon = C_0 \varepsilon^{(N-1)/(N-2)} & \text{if } N \geq 3 , \\ r_\star^\varepsilon = C_1 \exp(-C_0/\varepsilon) & \text{if } N = 2, \end{cases} \tag{8}$$

27

where $C_0 > 0$, $C_1 > 0$ are given constants. Then the solution u^ε of (5), (6), (7) satisfies[†]

$$\begin{cases} u_a^\varepsilon \rightharpoonup u_a & \text{in } H^1(\Omega_a) \ \text{weak}, \\ u_b^\varepsilon \rightharpoonup u_b & \text{in } H^1(\Omega_b) \ \text{weak}, \end{cases} \tag{9}$$

where the couple (u_a, u_b) is the solution of the following boundary value problem:

$$\begin{cases} -\Delta u_a + u_a = f_a & \text{in } \Omega_a , \\ -\Delta u_b + u_b = f_b & \text{in } \Omega_b , \end{cases} \tag{10}$$

$$\begin{cases} u_a = 0 & \text{on } (\partial\Omega)_a , \\ u_b = 0 & \text{on } (\partial\Omega)_b , \end{cases} \tag{11}$$

$$\begin{cases} \dfrac{\partial u_a}{\partial n_a} + \mu(u_a - u_b) = 0 & \text{on } \Gamma , \\[2mm] \dfrac{\partial u_b}{\partial n_b} + \mu(u_b - u_a) = 0 & \text{on } \Gamma . \end{cases} \tag{12}$$

Here $\partial u_a / \partial n_a = -\partial u_a / \partial x_N$ and $\partial u_b / \partial n_b = \partial u_b / \partial x_N$ denote the normal derivatives (according the outer normals n_a and n_b to Ω_a and Ω_b) and μ is the constant given by

$$\begin{cases} \mu = \dfrac{C_0^{N-2}}{2^{N+1}} \ \mathrm{Cap}_{\mathbb{R}^R}(\overline{\mathcal{K}}) & \text{if } N \ge 3, \\[3mm] \mu = \pi / (4C_0) & \text{if } N = 2, \end{cases} \tag{13}$$

[†] For any function v belonging to $L^1(\Omega)$ we define $v_a \in L^1(\Omega_a)$ and $v_b \in L^1(\Omega_b)$ as the restrictions of v to the upper and lower parts Ω_a and Ω_b. We use the same notation $w_a \in L^1(\Omega_a)$ and $w_b \in L^1(\Omega_b)$ for the restrictions of a function w belonging to $L^1(\Omega^\varepsilon)$.

where $\mathrm{Cap}_{\mathbb{R}^N}(\overline{\mathcal{H}})$ is the capacity (relative to $\mathbb{R}^N$) of the closure $\overline{\mathcal{H}}$ of the basic hole $\mathcal{H}$:

$$\mathrm{Cap}_{\mathbb{R}^N}(\overline{\mathcal{H}}) = \inf \int_{\mathbb{R}^N} |\nabla w|^2, \quad w \in \mathcal{D}(\mathbb{R}^N) \quad w = 1 \text{ on } \overline{\mathcal{H}}. \qquad (14)$$

Let us emphasize that the limit $u = (u_a, u_b)$ of u^ε is not continuous across Γ, but presents a jump which links the top and bottom normal derivatives: this is the asymptotic memory of the sieve.

If the holes are too small, i.e. if r^ε decreases more quickly than $r_\star^\varepsilon$ given by (8), there is nothing flowing through the sieve: one obtains two problems in Ω_a and Ω_b, which are completely uncoupled: (10), (11) and (12) with $\mu = 0$. Conversely, if the holes are too big, i.e. if r^ε is asymptotically bigger than $r_\star^\varepsilon$, there is no sieve effect: the limit function u has no jump across Γ (i.e. $u \in H^1(\Omega)$) and satisfies the equation in the whole Ω:

$$\begin{cases} -\Delta u + u = f \text{ in } \Omega, \\ u = 0 \qquad \text{on } \partial\Omega. \end{cases}$$

This corresponds to $\mu = +\infty$ in (12).

That μ appearing in (12) is expressed in terms of the capacity of the hole (see (13)) is a peculiarity due to the zero thickness of the sieve. For a generally thick sieve, the limit boundary condition on Γ is still given by (12), but μ is not expressed as **a** capacity; it can be expressed in terms of the solution of some special minimization problem.

Most of the techniques we use in the proof have their origin in an earlier joint work with Doina Cioranescu ([3], see in particular Example 2.9 and Section 3). This work was concerned with a Dirichlet problem which is (in some sense) complementary to the present Neumann problem. Finally we use in the proof a compactness lemma which perhaps is of some intrinsic interest (see [8]).

The problem of the Neumann sieve was brought to light by Enrique Sanchez-Palencia, who gave in [14] (Section 4) a formal asymptotic expansion of the solution u^ε. The main physical motivation of this type of problem is to study the flow of a fluid through a perforated wall (see, for example, J. Sanchez-Hubert and E. Sanchez-Palencia [12]). In this case, the boundary value problem is more complicated than ours, which is only a first academic study in this direction. We just point out that in the problem (5), (6) that we consider here, the unknown u^ε resembles the pressure (or the potential function of the velocity v^ε, if this is assumed to be curlfree) of an inviscid fluid flow through a perforated wall subject to the classical boundary condition $v^\varepsilon . n = 0$ of tangential motion on the wall (see [12], equations (2.2), (2.3), (2.6b) or (4.3), (4.4), (4.7)). This physical motivation is quite well adapted to the culinary context where this lecture takes place: let us just note that the sieve used to cook pasta is called "colabrodo" in Italian.

To conclude this résumé, let us emphasize that we consider here the Laplace equation (5) with the Neumann boundary condition (6) on a sieve. This is a very particular boundary value problem with holes: for other problems with different holes, see (among others): D. Cioranescu and J. Saint Jean Paulin [4] and J. L. Lions [9] Chapter 1 (the Neumann problem with distributed holes of size ε); D. Cioranescu and F. Murat [3] (the Dirichlet problem with distributed holes of size $\varepsilon^{N/(N-2)}$); J. L. Lions [9] Chapter 1 (the Dirichlet problem with distributed holes of size ε); J. L. Lions [9] Chapter 2 and E. Sanchez-Palencia [15] Chapter 7 and appendix (the Dirichlet boundary condition for the Stokes problem with distributed holes of size ε leading to Darcy's law for a flow in a porous medium); C. Conca [5] (the "Neumann" boundary condition for the Stokes problem with distributed holes of size ε); E. Sanchez-Palencia [13] and C. Conca [6] (Stokes flow through a perforated wall with the Dirichlet boundary condition with holes of size ε); etc.

The work we have presented here is confined to the Neumann problem (5), (6) on a sieve of null thickness and with a regular (periodic) distribution of holes. However, it can be extended in some directions: to the case of a sieve with thickness h^ε, possibly with non-cylindrical holes corresponding to a "stamped" sieve; to the case of a semipermeable sieve; to the case of a general (non-periodic) distribution of holes on the hyperplane $x_N = 0$; to the case of a pile of sieves, etc. (for all these extensions, see our selected works [1], [2], [7], [11]).

Finally let us note that in the problem (5), (6), (7) that we consider here, the main role is played by the Neumann boundary condition (6) on Γ^ε. The boundary condition (7) on the outer boundary could be replaced by any variational condition. In the same way, the zero-order term u^ε is just used in (5) for the sake of simplicity and could be suppressed.

REFERENCES

1. Hedy Attouch, Alain Damlamian, François Murat and Colette Picard, The Neumann Sieve (to appear).

2. Hedy Attouch and Colette Picard, Comportement limite de problèmes variationnels dans un domaine contenant une passoire de Neumann quelconque. Rend. Sem. Mat. Univers. Politecn. Torino (to appear).

3. Doina Cioranescu and François Murat, Un terme étrange venu d'ailleurs. Nonlinear Partial Differential Equations and Their Applications, Collège de France Seminar, II and III, ed. H. Brézis and J. L. Lions, Research Notes in Mathematics 60 and 70, Pitman, London, (1982), 98-138 and 154-178.

4. Doina Cioranescu and Jeannine Saint Jean Paulin, Homogenization in open sets with holes. J. Math. Anal. Appl., 71 (1979) 590-607.

5. Carlos Conca, On the application of the homogenization theory to a class of problems arising in fluid mechanics. J. Math. Pures et Appl. (to appear).

6. Carlos Conca, Comportement asymptotique d'un fluide traversant une passoire (to appear).

7. Alain Damlamian, Le problème de la passoire de Neumann. Rend. Sem. Mat. Univers. Politecn. Torino (to appear).

8. Lars Hedberg and François Murat, Les intervalles de $W^{-s,p}$ sont compacts (to appear).

9. Jacques-Louis Lions, Some Methods in the Mathematical Analysis of Systems and Their Control. Science Press, Beijing, and Gordon and Breach, New York (1981).

10. François Murat, L'injection du cone positif de H^{-1} dans $W^{-1,q}$ est compacte pour tout $q < 2$. J. Math. Pures et Appl. , 60 (1981) 309-322.

11. Colette Picard, Analyse limite d'équations variationelles dans un domaine contenant une grille. In Thèse d'Etat, Université de Paris-Sud (Orsay) (1984).

12. Jacqueline Sanchez-Hubert and Enrique Sanchez-Palencia, Acoustic fluid flow through holes and permeability of perforated walls. J. Math. Anal. and Appl. , 87 (1982) 427-453.

13. Enrique Sanchez-Palencia, Non-homogeneous Media and Vibration Theory. Lecture Notes in Physics 127, Springer, Berlin (1980).

14. Enrique Sanchez-Palencia, Boundary value problems in domains containing perforated walls. Nonlinear Partial Differential Equations and Their Applications, Collège de France Seminar, III, ed. H. Brézis and J. L. Lions, Research Notes in Mathematics 70, Pitman, London (1982) 309-325.

15. Enrique Sanchez-Palencia, Un problème d'écoulement lent d'un fluide visqueux incompressible au travers d'une paroi finement perforée. Cours de l'Ecole d'été d'Analyse Numérique CEA-EDF-INFIA sur l'homogénéisation (Bréau sans Nappe, juillet 1983). Collection de la Direction des Etudes et Recherches d'Electricité de France, Eyrolles, Paris (to appear).

François Murat
Analyse Numérique
Tour 55-65, 5e étage
Université de Paris VI
4 Pl. Jussieu
75230 Paris Cedex 05
France

H BREZIS
Large harmonic maps in two dimensions

In this lecture I report about a joint work with J. M. Coron - see $[4]$.

Let

$$\Omega = \{(x, y) \in \mathbb{R}^2 ; \quad x^2 + y^2 < 1\}$$

and

$$S^2 = \{(x, y, z) \in \mathbb{R}^3 ; \quad x^2 + y^2 + z^2 = 1\}.$$

We are concerned with the following question: find $u : \Omega \to \mathbb{R}^3$ satisfying

$$\begin{cases} -\Delta u = u |\nabla u|^2 & \text{in } \Omega, \\ u(x, y) \in S^2 & \text{in } \Omega, \\ u = \gamma & \text{in } \partial\Omega, \end{cases} \tag{1}$$

where $\gamma : \partial\Omega \to S^2$ is given.

The solutions of (1) are <u>harmonic maps</u> with the prescribed boundary condition $u = \gamma$ on $\partial\Omega$. Problem (1) has a variational structure. It is easy to see that the solutions of (1) correspond to the critical points of the functional

$$E(u) = \int_\Omega |\nabla u|^2$$

subject to the constraint

$$u \in \mathcal{E} = \{u \in H^1(\Omega; \mathbb{R}^3) ; \ u \in S^2 \text{ a. e. on } \Omega, \ u = \gamma \text{ on } \partial\Omega\}.$$

In what follows we assume that $\mathcal{E} \neq \emptyset$, that is, γ is the trace on $\partial\Omega$ of some function $u \in H^1(\Omega; \mathbb{R}^3)$ with $u \in S^2$ a. e. on Ω.

<u>Remark 1</u>　Most of the properties discussed below still hold if we deal with the critical points of E subject to the constraint

33

$$u \in \mathcal{E}' = \{ u \in H^1(\Omega; \mathbb{R}^3) \, ; \; u \in M \text{ a.e. on } \Omega, \; u = \gamma \text{ on } \partial\Omega \}$$

where $M \subset \mathbb{R}^3$ is a Riemannian surface homeomorphic to S^2 and $\gamma \in M$ a.e. on $\partial\Omega$.

Remark 2   Here we are concerned only with the question of <u>existence of weak solutions</u> – $u \in H^1$ – and not with the question of <u>regularity</u>. It is an interesting open problem to determine whether any H^1 solution of (1) is smooth. We briefly recall some related known facts:

(a) Some elliptic systems which resemble (1) do admit weak <u>nonsmooth</u> solutions; see e.g. Giaquinta [7] and Hildebrandt [9].

(b) If u is an H^1 solution of (1) and moreover u is continuous on Ω, then u is smooth; see e.g. [7] or [9].

(c) If $\underline{u} \in H^1$ is an <u>absolute minimum</u>, that is, $\underline{u} \in \mathcal{E}$ and $E(\underline{u}) \le E(v)$ $\forall v \in \mathcal{E}$, then $\underline{u}$ is smooth (this result is due to Morrey).

(d) An H^1 solution of (1) which is smooth on $\Omega \setminus \{ x_0 \}$ is smooth on Ω (this result is due to Sacks-Uhlenbeck [14]).

It is quite trivial that there is some $\underline{u} \in \mathcal{E}$ which is an absolute minimum of E, that is

$$E(\underline{u}) \le E(v) \qquad \forall v \in \mathcal{E}$$

(indeed, consider a minimizing sequence and pass to the limit with the help of the lower semicontinuity for the weak H^1 topology).

Our main result is the following:

Theorem 1   Suppose that γ is not a constant. Then there exist at least two distinct solutions of (1). More precisely, there is some $\bar{u} \neq \underline{u}$ which is a local minimum of E on $\mathcal{E}$.

Remark 3   When $\gamma = C$ is a constant, it is known (see Lemaire [11]) that $u \equiv C$ is the only solution of (1); such a result is "loosely" related to the results of Pohozaev [13] and Wente [18] respectively for the problems $-\Delta u = u^{(N+2)/(N-2)}$

on $\Omega \subset \mathbb{R}^N$ and $\Delta u = 2u_x \wedge u_y$ on Ω (Ω as above).

<u>Remark 4</u>   Theorem 1 answers a question raised by Giaquinta-Hildebrandt in [8]. They had observed in the special case where

$$\gamma(x, y) = \begin{pmatrix} Rx \\ Ry \\ \sqrt{(1 - R^2)} \end{pmatrix} \qquad \text{with } 0 < R \le 1 \tag{2}$$

that there are two distinct solutions of (1):

(a)   the "small bubble" spanned by $\gamma(\partial\Omega)$:

$$\underline{u}(x, y) = \frac{2\lambda}{\lambda^2 + r^2} \begin{pmatrix} x \\ y \\ \lambda \end{pmatrix} + \begin{pmatrix} 0 \\ 0 \\ -1 \end{pmatrix}$$

where $r^2 = x^2 + y^2$ and $\lambda = \frac{1}{R} + \sqrt{(\frac{1}{R^2} - 1)}$,

(b)   the "large bubble" spanned by $\gamma(\partial\Omega)$:

$$\bar{u}(x, y) = \frac{2\mu}{\mu^2 + r^2} \begin{pmatrix} x \\ y \\ \mu \end{pmatrix} + \begin{pmatrix} 0 \\ 0 \\ 1 \end{pmatrix}$$

where $r^2 = x^2 + y^2$ and $\mu = \frac{1}{R} - \sqrt{(\frac{1}{R^2} - 1)}$

($\underline{u}$ and $\bar{u}$ are, modulo a dilation, stereographic projections from the north and the south poles: see Figure 1).

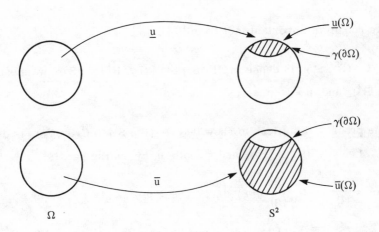

Figure 1

35

Incidentally, it is not known, and it would be interesting to determine, whether $\underline{u}$ and $\bar{u}$ are the only solutions of (1) corresponding to γ given by (2); see, however, Theorem 2.

This example suggests some analogy with the problem of finding surfaces of constant mean curvature spanned by a given curve (Rellich's conjecture), for which Coron and myself [3] have proved the existence of two distinct solutions.

Remark 5   A result comparable to Theorem 1 has been obtained independently by Jost [10].

The proof of Theorem 1 involves a splitting of $\mathscr{E}$ into (connected) components classified by their degree. Since we deal with H^1 mappings, they need not be continuous and thus their degree is not defined in a standard way. However, following Schoen and Uhlenbeck [15] we observe that H^1 maps from S^2 into S^2 do have a well defined degree.

Indeed, suppose first that $\phi \in C^1(S^2; S^2)$ then $\deg \phi$ is defined and there is even an analytic expression for the degree

$$\deg \phi = \frac{1}{4\pi} \int_{S^2} J_\phi \tag{3}$$

where J_ϕ is the Jacobian of ϕ (see, e.g., L. Nirenberg [12]). Note that J_ϕ is quadratic in the derivatives of ϕ and thus $\int_{S^2} J_\phi$ makes sense when ϕ belongs to H^1.

Next we recall an important fact from [15].

Lemma 1 $C^\infty(S^2; S^2)$ is dense in $H^1(S^2; S^2)$ (by $H^1(S^2; S^2)$ we mean $u \in H^1(S^2; \mathbb{R}^3)$ and $u \in S^2$ a.e.).

Sketch of the proof   It suffices to show that $H^1(S^2; S^2) \cap C(S^2; S^2)$ is dense in $H^1(S^2; S^2)$. Let $u \in H^1(S^2; S^2)$ and smooth u by averaging

$$u_\varepsilon(q) = \frac{1}{|B_\varepsilon|} \int_{B_\varepsilon(q)} u(\xi)\, d\xi, \qquad \varepsilon > 0$$

so that $u_\varepsilon \in H^1(S^2, \mathbb{R}^3) \cap C(S^2, \mathbb{R}^3)$ and $u_\varepsilon \to u$ in H^1 as $\varepsilon \to 0$. We claim that

$$\text{dist}(u_\varepsilon(q), S^2) \xrightarrow[\varepsilon \to 0]{} 0 \quad \text{uniformly for } q \in S^2. \tag{4}$$

This suffices to conclude, since we may then consider $\bar{u}_\varepsilon = \text{Proj}_{S^2} u_\varepsilon$.

In order to prove (4), we note that by Poincaré's inequality we have

$$\int_{B_\varepsilon(q)} |u(\xi) - u_\varepsilon(q)| \, d\xi \le C |B_\varepsilon|^{\frac{1}{2}} \int_{B_\varepsilon(q)} |\nabla u|$$

and thus

$$\int_{B_\varepsilon(q)} |u(\xi) - u_\varepsilon(q)| \, d\xi \le C |B_\varepsilon| (\int_{B_\varepsilon(q)} |\nabla u|^2)^{\frac{1}{2}}.$$

Therefore

$$\text{dist}(u_\varepsilon(q), S) \le C (\int_{B_\varepsilon(q)} |\nabla u|^2)^{\frac{1}{2}} \xrightarrow[\varepsilon \to 0]{} 0$$

since $u \in H^1$.

<u>Remark 6</u>   The same argument as above shows that

$$C^\infty(\Omega; S) \text{ is dense in } W^{1,p}(\Omega; S)$$

where Ω is a smooth manifold (with or without boundary) of dimension N, S is a smooth manifold without boundary (of any dimension) and $p \ge N$. However, if $p < N$ it may happen that $C^\infty(\Omega, S)$ is not dense in $W^{1,p}(\Omega, S)$. Let, for example, Ω be the unit ball in $\mathbb{R}^3$ and $S = S^2$. Then

$$u(x) = \frac{x}{|x|} \in H^1(\Omega; S^2),$$

but $u \notin \overline{C^\infty(\Omega; S^2)}^{H^1}$. Indeed, suppose by contradiction that $u_n \to u$ in H^1 and $u_n \in C^\infty(\Omega; S^2)$. It follows that $u_n\big|_{rS^2} \to u\big|_{rS^2}$ in H^1 for a.e. r. Set $\phi_n(x) = u_n(rx)$, $x \in S^2$ for such an r, so that $\phi_n \in C^\infty(S^2; S^2)$. Then $\phi_n \to \text{Id}$ in H^1 and, moreover, $\deg \phi_n = 0$ (because ϕ_n is homotopic to the constant map $u_n(0)$). This is absurd since

$$0 = \deg \phi_n = \frac{1}{4\pi} \int_{S^2} J_{\phi_n} \rightarrow \frac{1}{4\pi} \int_{S^2} J_{Id} = 1.$$

<u>Remark 7</u> (Boutet de Monvel, personal communication). The discussion above suggests a theory of degree for mappings which are merely BMO with small bounded oscillation, for example such that

$$\lim_{|Q| \to 0} \frac{1}{|Q|} \int_Q |\phi - \phi_Q| = 0 \quad \text{where} \quad \phi_Q = \frac{1}{|Q|} \int_Q \phi .$$

We are now ready to split $\mathcal{E}$ into components. Given u_1, $u_2 \in \mathcal{E}$, we transport them and glue them on to S^2 in the following way. Let π_+ (resp π_-) denote the stereographic projection from S^2 into $\mathbb{R}^2$ with vertex at the south pole (resp. north pole). Note that π_+ maps S^2_+ (the northern hemisphere) onto Ω and π_- maps S^2_- (the southern hemisphere) onto Ω (Figure 2). We set

$$\phi(p) = \begin{cases} u_1(\pi_+(p)) & \text{if } p \in S^2_+ \\ u_2(\pi_-(p)) & \text{if } p \in S^2_- \end{cases}$$

Figure 2

Note that the two mappings "glue well" since $u_1 = u_2 = \gamma$ on $\partial \Omega$, and therefore ϕ belongs to $H^1(S^2; S^2)$. Thus we may define its degree. This degree can be computed in terms of u_1, u_2 and we find

$$\deg \phi = \frac{1}{4\pi}(Q(u_1) - Q(u_2))$$

where $Q(u) = \int_\Omega u(u_x \wedge u_y)$.

We fix arbitrarily some reference element in $\mathcal{E}$; it is convenient to use $\underline{u}$ and we set for $k \in \mathbb{Z}$

$$\mathcal{E}_k = \{u \in \mathcal{E}; \frac{1}{4\pi} (Q(u) - Q(\underline{u})) = k\}.$$

Clearly $\mathcal{E} = \bigcup_{k \in \mathbb{Z}} \mathcal{E}_k$ and each $\mathcal{E}_k$ is both open and closed; moreover $\mathcal{E}_k \neq \emptyset \ \forall k$, and $\underline{u} \in \mathcal{E}_0$.

In order to find critical points of E on $\mathcal{E}$ it is quite tempting to consider

$$\underset{\mathcal{E}_k}{\text{Inf}} \ E.$$

If the infimum is achieved then we obtain a critical point of E and thus a solution of (1). However, there is a serious difficulty in trying to carry out this program: the sets $\mathcal{E}_k$ are closed for the strong H^1 topology, but they are not closed for the weak H^1 topology. Hence if (u^j) is a minimizing sequence in $\mathcal{E}_k$, (u^j) is bounded in H^1 and so $u^j \rightharpoonup u$ weakly in H^1. By lower semicontinuity we have

$$E(u) \leq \underset{\mathcal{E}_k}{\text{Inf}} \ E.$$

However, it may well happen that $u \notin \mathcal{E}_k$; this does indeed happen, for example, if $\gamma = C$ and $k \neq 0$ (see Remark 3).

A similar difficulty occurs in the Yamabe problem (see $[1]$ and $[6]$). Consider, for example, the minimization problem

$$\underset{\substack{u \in H_0^1(\Omega) \\ \|u\|_{L^{2^\star}} = 1}}{\text{Inf}} \int_\Omega |\nabla u|^2 + a(x) u^2$$

where $\Omega \subset \mathbb{R}^N$ is a smooth domain and $2^\star = \frac{2N}{N-2}$. A major difficulty arises from the fact that the set $\{u \in H_0^1(\Omega); \|u\|_{L^{2^\star}} = 1\}$ is closed for the strong H^1

topology but not for the weak H^1 topology.

Again, a similar obstacle occurs in [3] (Rellich's conjecture) where one deals with

$$\underset{\substack{u \in H_0^1(\Omega; \mathbb{R}^3) \\ Q(u) = 1}}{\text{Inf}} \quad \{ \int_\Omega |\nabla u|^2 + 4 \int_\Omega \underline{u} \; u_x \wedge u_y \}$$

and the set $\{u \in H_0^1 ; \; Q(u) = 1\}$ is closed for the strong, but not for the weak H^1 topology.

Sketch of the proof of Theorem 1   The proof is divided into two steps:

Step 1 $\underset{\mathcal{E}_1 \cup \mathcal{E}_{-1}}{\text{Inf}} \; E < \underset{\mathcal{E}}{\text{Inf}} \; E + 8\Pi,$

Step 2 $\underset{\mathcal{E}_1 \cup \mathcal{E}_{-1}}{\text{Inf}} \; E$ is achieved.

The idea underlying this approach is that below some "magic" levels a form of compactness appears in the problem - the same strategy is used in [1], [3], [4], [6].

Step 1   It suffices to find a function $v \in \mathcal{E}$ such that

$$\frac{1}{4\pi} \left| Q(v) - Q(\underline{u}) \right| = 1 \tag{5}$$

$$\int_\Omega |\nabla v|^2 < \int |\nabla \underline{u}|^2 + 8\Pi . \tag{6}$$

The construction of v is explicit and rather technical. One fixes a point $(x_0, y_0) \in \Omega$ such that $\nabla \underline{u}(x_0, y_0) \neq 0$ (the assumption γ non-constant enters here). Set

$$D_\varepsilon = \{(x, y) ; \; (x-x_0)^2 + (y-y_0)^2 < \varepsilon^2 \}.$$

One chooses v^ε such that:

(a) $v^\varepsilon = \underline{u}$ in $\Omega \setminus D_{2\varepsilon}$

40

(b) $v^{\varepsilon}(x, y) = \dfrac{2\lambda}{\lambda^2 + r^2} \begin{pmatrix} x-x_0 \\ y-y_0 \\ -\lambda \end{pmatrix} + \begin{pmatrix} 0 \\ 0 \\ 1 \end{pmatrix}$ in D_{ε}

where $\lambda = c\varepsilon^2$, $r^2 = (x-x_0)^2 + (y-y_0)^2$ and c is a constant to be fixed.

A careful expansion as $\varepsilon \to 0$ shows that

$$\frac{1}{4\pi} \left| Q(v^{\varepsilon}) - Q(\underline{u}) \right| = 1$$

and

$$\int \left| \nabla v^{\varepsilon} \right|^2 = \int \left| \nabla \underline{u} \right|^2 + 8\Pi - \alpha\varepsilon^2 + o(\varepsilon^2)$$

with $\alpha > 0$ (for an appropriate choice of c).

Step 2   Suppose, for example, that

$$J = \underset{\mathcal{E}_1}{\text{Inf }} E < \int_{\Omega} \left| \nabla \underline{u} \right|^2 + 8\Pi.$$

We shall prove that Inf E is achieved. Let (u^j) be a minimizing sequence and
 $\mathcal{E}_1$
so we have

$$\int_{\Omega} \left| \nabla u^j \right|^2 = J + o(1) \tag{7}$$

$$Q(u^j) - Q(\underline{u}) = 4\Pi. \tag{8}$$

We may assume that

$u^j \rightharpoonup \bar{u}$ weakly in H^1

$u^j \to \bar{u}$ a. e. on Ω.

Clearly we have

$$\int \left| \nabla \bar{u} \right|^2 \leq J$$

and the main difficulty is to prove that $\bar{u} \in \mathcal{E}_1$, that is

$$Q(\bar{u}) - Q(\underline{u}) = 4\Pi . \qquad (9)$$

Set $v^j = u^j - \bar{u}$ so that $v^j \to 0$ weakly in H_0^1 and

$$\int |\nabla \bar{u}|^2 + \int |\nabla v^j|^2 = J + o(1) . \qquad (10)$$

On the other hand, we have

$$
\begin{aligned}
Q(u^j) &= \int u^j \cdot (\bar{u}_x + v^j_x) \wedge (\bar{u}_y + v^j_y) \\
&= \int \bar{u} \cdot \bar{u}_x \wedge \bar{u}_y + o(1) + \int u^j \cdot v^j_x \wedge v^j_y
\end{aligned}
$$

(since products of the form $u^j \wedge \bar{u}_x$ and $u^j \wedge \bar{u}_y$ converge strongly in L^2 by dominated convergence). Thus we obtain

$$\left| Q(u^j) - Q(\bar{u}) \right| \le \tfrac{1}{2} \int |\nabla v^j|^2 + o(1) . \qquad (11)$$

Combining (8) and (11), we find

$$\left| Q(\underline{u}) - Q(\bar{u}) + 4\Pi \right| \le \tfrac{1}{2} \int |\nabla v^j|^2 + o(1)$$

(by (10))
$$\le \tfrac{1}{2} \left[J - \int |\nabla \bar{u}|^2 \right] + o(1)$$

$$\le \tfrac{1}{2} \left[J - \int |\nabla \underline{u}|^2 \right] + o(1).$$

Using Step 1, we see that

$$\left| Q(\underline{u}) - Q(\bar{u}) + 4\pi \right| < 4\pi$$

and thus

$$\left| \frac{1}{4\pi} (Q(\underline{u}) - Q(\bar{u})) + 1 \right| < 1.$$

Since $\frac{1}{4\pi} (Q(\underline{u}) - Q(\bar{u})) \in \mathbb{Z}$, we conclude that $\frac{1}{4\pi} (Q(\underline{u}) - Q(\bar{u})) = -1$, that is $\bar{u} \in \mathcal{E}_1$.

We have an additional result in the special case where γ is given by (2) with $0 < R < 1$.

Theorem 2   Suppose that γ is given by (2) with $0 < R < 1$. Then we have:

(a) Min E is achieved only at u,
     $\mathcal{E}$

(b) $\bar{u} \in \mathcal{E}_{-1}$ and Min E is achieved only at $\bar{u}$,
     $\mathcal{E}_{-1}$

(c) Inf E is not achieved when $k \neq 0$ and $k \neq -1$.
     $\mathcal{E}_k$

Remark 8   Theorem 2 does not rule out the existence of other critical points of E on $\mathcal{E}_0$, $\mathcal{E}_{-1}$ or $\mathcal{E}_k$ ($k \neq 0$, $k \neq -1$). This is an interesting open problem.

For the proof of Theorem 2 we refer to [4]; however, we mention the following lemma which plays a crucial role in the proof of Theorem 2.

Lemma 2   Suppose that $\phi \in L^\infty(\mathbb{R}^2, S^2)$ with $\phi_x, \phi_y \in L^2(\mathbb{R}^2, \mathbb{R}^3)$ satisfies

$$2 \int_{\mathbb{R}^2} \phi \cdot \phi_x \wedge \phi_y = \int_{\mathbb{R}^2} |\nabla \phi|^2 . \tag{12}$$

Then ϕ is (real) analytic and, more precisely, ϕ is given by

$$\phi(\zeta) = -\Pi \left(\frac{P(\zeta)}{Q(\zeta)} \right) , \qquad \zeta = (x, y) = x + iy$$

where Π is the stereographic projection from the north pole and P, Q are two polynomials.

Sketch of the proof of Lemma 2   Since $|\phi| = 1$ it follows that $\phi \cdot \phi_x = \phi \cdot \phi_y = 0$ and therefore

$$\phi_x \wedge \phi_y = \lambda \phi \tag{13}$$

where λ is some function in $L^1(\mathbb{R}^2; \mathbb{R})$. We have

$$|\lambda| = |\phi_x \wedge \phi_y| \leq \tfrac{1}{2} |\nabla \phi|^2 . \tag{14}$$

Combining (12), (13), (14) we see that

$$\lambda = \tfrac{1}{2} |\nabla \phi|^2 \tag{15}$$

$$\phi_x^2 - \phi_y^2 = \phi_x \cdot \phi_y = 0, \tag{16}$$

that is, ϕ is conformal.

On the other hand we clearly have

$$\int \zeta \ \zeta_x \wedge \zeta_y \leq \tfrac{1}{2} \int |\zeta| \ |\nabla \zeta|^2 \tag{17}$$

$$\forall \zeta \in L^\infty (\mathbb{R}^2 ; \mathbb{R}^3) \quad \text{with} \quad \zeta_x, \ \zeta_y \in L^2 (\mathbb{R}^2 ; \mathbb{R}^3).$$

Inserting in (17)

$$\zeta = \phi + t\psi \qquad \text{with} \ \psi \in \mathcal{D}(\mathbb{R}^2 ; \mathbb{R}^3)$$

we find (using (12))

$$3t \int \psi \cdot \phi_x \wedge \phi_y \leq \frac{t}{2} \int \phi \cdot \psi |\nabla \phi|^2 + 2 \nabla \phi \nabla \phi + o(t).$$

But (13) and (15) imply that

$$\psi \cdot \phi_x \wedge \phi_y = \lambda \phi \cdot \psi = \tfrac{1}{2}\phi \cdot \psi |\nabla \psi|^2.$$

Therefore we are left with

$$2 \int \psi \cdot \phi_x \wedge \phi_y = \int \nabla \phi \nabla \psi,$$

that is,

$$-\Delta \phi = 2 \phi_x \wedge \phi_y \qquad \text{in } \mathcal{D}'. \tag{18}$$

We deduce from a result of Wente [18] that ϕ is smooth and even (real) analytic. Combining (13), (15) and (18) we see that

$$-\Delta \phi = \phi |\nabla \phi|^2 \qquad \text{on } \mathbb{R}^2,$$

that is, ϕ is harmonic.

Finally one shows (as in [5]) that $\phi \circ \Pi^{-1}$ is smooth on S^2, thus it is a smooth harmonic map from S^2 into S^2. Since all such maps are known (see [16] or $[11]$) we deduce that $\phi = -\pi\left(\dfrac{P}{Q}\right)$.

<u>Remark 9</u> Benci and Coron $[2]$ have recently considered the following problem: find $u : \Omega \to \mathbb{R}^{n+1}$ satisfying

$$
\begin{cases}
-\Delta u = u\,|\nabla u|^2 & \text{in } \Omega \ (\Omega \text{ as before})\,, \\
u(x,\,y) \in S^n & \text{in } \Omega \\
u = \gamma & \text{in } \partial\Omega\,.
\end{cases}
\tag{19}
$$

They prove that if γ is non-constant and $C^{2,\delta}$ $(0 < \delta \le 1)$, then there exist at least two solutions of (19). The argument is completely different from the one described here since $\mathcal{E}$ is <u>connected when</u> $n \ge 3$.

REFERENCES

1. Th. Aubin, Equations différentielles non linéaires et problème de Yamabe concernant la coubure scalaire, <u>J. Math. Pures et Appl.</u> 55 (1976) 269-296.

2. V. Benci and J. M. Coron, The Dirichlet problem for harmonic maps from the disk into the euclidean n-sphere, <u>Ann. IHP Analyse Nonlinéaire</u> (to appear).

3. H. Brezis and J. M. Coron, Multiple solutions of H-Systems and Rellich's conjecture, <u>Comm. Pure Appl. Math.</u> 37 (1984) 149-187.

4. H. Brezis and J. M. Coron, Large solutions for harmonic maps in two dimensions, <u>Comm. Math. Phys.</u> 92 (1983) 203-215.

5. H. Brezis and J. M. Coron, Convergence of solutions of H-Systems or how to blow bubbles, <u>Archive Rat. Mech. Anal.</u> (to appear).

6. H. Brezis and L. Nirenberg, Positive solutions of nonlinear elliptic equations involving critical Sobolev exponents, <u>Comm. Pure Appl. Math.</u> 36 (1983) 437-477.

7. M. Giaquinta, <u>Multiple Integrals in the Calculus of Variations and Nonlinear Elliptic Systems</u>, Princeton Univ. Press (1983).

8. M. Giaquinta and S. Hildebrandt, A priori estimates for harmonic mappings, <u>J. Reine Angew. Math.</u> 336 (1982) 124-164.

9. S. Hildebrandt, Nonlinear elliptic systems and harmonic mappings, in <u>Proc. Beijing Symp. on Diff. Geom. and Diff. Eq.</u>, Science Press Beijing (1982).

10.   J. Jost, The Dirichlet problem for harmonic maps from a surface with boundary onto a 2-sphere with non constant boundary values, <u>J. Diff. Geom.</u> (to appear).

11.   L. Lemaire, Applications harmonique de surfaces riemaniennes, <u>J. Diff. Geom.</u> <u>13</u> (1978) 51-68.

12.   L. Nirenberg, <u>Topics in Nonlinear Functional Analysis</u>, NYU Lecture Notes 1973-74.

13.   S. Pohozaev, Eigenfunctions of the equation $\Delta u + \lambda f(u) = 0$, <u>Soviet Math. Doklady</u> <u>6</u> (1965) 1408-1411.

14.   J. Sacks and K. Uhlenbeck, The existence of minimal immersions of 2-spheres, <u>Annals of Math.</u> <u>113</u> (1981) 1-24.

15.   R. Schoen and K. Uhlenbeck, Boundary regularity and the Dirichlet problem for harmonic maps, <u>J. Diff. Geom.</u> <u>18</u> (1983) 253-268.

16.   G. Springer, <u>Introduction to Riemann Surfaces</u>, Addison-Wesley (1957).

17.   H. Wente, An existence theorem for surfaces of constant mean curvature, <u>J. Math. Anal. Applic.</u> <u>26</u> (1969) 318-344.

18.   H. Wente, The differential equation $\Delta x = 2H x_u \wedge x_v$ with vanishing boundary values, <u>Proc. AMS</u>, <u>50</u> (1975) 131-137.

Haim Brezis
Analyse Numérique
Tour 55-65, 5e étage
Université de Paris VI
4 Pl. Jussieu
75230 Paris Cedex 05
France

P-L LIONS
Symmetries and the concentration–compactness method

INTRODUCTION

In $[7]$, $[8]$ we introduced a general method - called the concentration-compactness method - to solve minimization problems (in function spaces) with possible loss of compactness. This lack of compactness comes from the invariance of $\mathbb{R}^N$ (say) by the noncompact groups of translations or dilations. This difficulty is typical of various problems of mathematical physics or differential geometry (see P. L. Lions $[8]$, H. Brézis $[3]$, K. Uhlenbeck $[14]$ ...).

The simple idea behind our general method is to compare our original minimization problem with a series of other minimization problems which are obtained by considering all the possible losses of compactness which may occur. Then, the original minimization problem is solvable if and only if some strict subadditivity condition holds. After a brief illustration of this method in some examples, we consider situations when the problem is invariant by a group of orthogonal transformations. Then, restriction of the minimization to invariant test functions modifies the possible loss of compactness and may even possibly eliminate noncompactness. The fact that symmetries may help to solve such minimization problems was used in W. Strauss $[12]$, H. Berestycki and P. L. Lions $[5]$, P. L. Lions $[9]$, $[10]$. We will show below in some examples that the concentration-compactness method is quite efficient in explaining such phenomena and we refer the reader to $[8]$ for a general presentation. Let us also mention that the arguments in $[8]$ that we recall below are closely related to the work of C. V. Coffman and Markus $[6]$.

As we said before, we will consider only a simple example. We wish to obtain a solution of

$$-\Delta u + a(x) u = u^p \quad \text{in } \Omega, \quad u > 0 \quad \text{in } \Omega, \quad u \in H_0^1(\Omega) \tag{1}$$

where $a(x)$ (to simplify) satisfies at least

$$a(x) \in C_b(\mathbb{R}^N), \quad a(x) \to a^\infty \quad \text{as} \quad |x| \to \infty. \tag{2}$$

Here and below, Ω is a smooth open set in $\mathbb{R}^N$ possibly unbounded. Using the homogeneity of the nonlinearity u^p, it is clear that a minimum of

$$I = \text{Inf} \left\{ \int_\Omega |\nabla u|^2 + au^2 \, dx \, / u \in H_0^1, \, \int_\Omega |u|^{p+1} dx = 1 \right\} \tag{3}$$

yields a solution of (1) (multiply u by a convenient positive constant). It is obvious that $I = 0$ (and there is no minimum) if $p > \dfrac{N+2}{N-2}$ and thus we will be interested in p lying in $(1, \dfrac{N+2}{N-2}]$ (or $(1, \infty)$ if $N \le 2$).

Observe, for example, that if $a(x) \equiv a^\infty$ on $\mathbb{R}^N$ and $\Omega = \mathbb{R}^N$ then problem (3) is translation invariant, while if $a \equiv 0$ problem (3) is dilation invariant or, more precisely, invariant under the scale change $(u \to t^{-(N-2)/2} u(\frac{\cdot}{t}))$ for $t > 0$.

Let us also remark that if $p < \dfrac{N+2}{N-2}$, the problem is "locally compact" in the sense that if Ω is bounded (and the operator $-\Delta + a$ is positive on Ω) then (3) is easy to solve by standard arguments. This is of course due to the well-known fact that the embedding $H_0^1(\Omega) \hookrightarrow L^{p+1}(\Omega)$ is compact if Ω is bounded and $p < \dfrac{N+2}{N-2}$. Again, one can see the lack of compactness due to translations or dilations by considering the case when $\Omega = \mathbb{R}^N$ and $p < \dfrac{N+2}{N-2}$, or $p = \dfrac{N+2}{N-2}$: indeed, in both cases the embedding $H_0^1(\Omega) \hookrightarrow L^{p+1}(\Omega)$ is no longer compact.

In Section 1 below, we consider the locally compact case (i.e. $p < \dfrac{N+2}{N-2}$) while Section 2 is devoted to the limit case (i.e. $p = \dfrac{N+2}{N-2}$). In both sections we first apply the concentration-compactness arguments to problem (3), and then we consider the cases when Ω, $a(x)$ are invariant by a group G of orthogonal transformations. In this case, one may still obtain a solution of (1) by considering

$$I_G = \text{Inf} \left\{ \int_\Omega |\nabla u|^2 + au^2 \, dx \, / u \in H_0^1(\Omega), \, \int_\Omega |u|^{p+1} dx = 1, \right.$$

$$\left. u \text{ is } G\text{-invariant} \right\} \tag{4}$$

where one says that u is G-invariant if $u(gx) = u(x)$ $x \in \Omega$, $g \in G$. In both sections, we will analyse (4) for the above values of p.

1.   THE LOCALLY COMPACT CASE

We consider here the case $1 < p < \dfrac{N+2}{N-2}$ if $N \geq 3$, $1 < p < \infty$ if $N \leq 2$. We will assume that a, Ω satisfy (to simplify):

$$\forall R < \infty, \ \exists y \in \Omega, \ B(y, R) \subset \Omega \tag{5}$$

$$\exists \nu > 0, \ \forall \phi \in H_0^1(\Omega), \ \int_\Omega |\nabla \phi|^2 + a\phi^2 dx \geq \nu \|\phi\|^2_{H_0^1} \tag{6}$$

To fix the ideas, one can think of the very special case when $\Omega = \mathbb{R}^N$. It is then clear enough that one has to avoid, when minimizing (3), minimizing sequences "slipping to infinite". The idea (in the simple homogeneous problem (3)) is then to compute the effect of slipping to infinity, i.e., let $\phi \in \mathcal{D}(\mathbb{R}^N)$, and consider

$$\lim_{|y| \to \infty} \mathcal{E}(\phi(\cdot + y)) = \mathcal{E}^\infty(\phi)$$

where $\mathcal{E}(v) = \int_\Omega |\nabla v|^2 + av^2 dx$. In view of (2), we have

$$\mathcal{E}^\infty(\phi) = \int_{\mathbb{R}^N} |\nabla \phi|^2 + a^\infty \phi^2 dx . \tag{7}$$

We then set

$$I^\infty = \text{Inf} \left\{ \int_{\mathbb{R}^N} |\nabla u|^2 + a^\infty u^2 dx / u \in H^1(\mathbb{R}^N), \ \int_{\mathbb{R}^N} |u|^{p+1} dx = 1 \right\}. \tag{8}$$

A small extension of the above argument shows that we always have $I \leq I^\infty$ and that if $I = I^\infty$ one can build minimizing sequences which are not compact in L^2 (or H_0^1, or L^{p+1} ...). In fact, the concentration compactness principle yields:

<u>Proposition 1.1</u>   We assume (2), (5), (6). Then, the inequality

$$I < I^\infty \tag{9}$$

49

is necessary and sufficient for the compactness of all minimizing sequences in H^1.

In all cases, any minimizing sequence $(u_n)_n$ is relatively compact in H^1 up to a translation, i.e., there exists $(y_n)_n$ in $\mathbb{R}^N$ such that $u_n(\cdot + y_n)$ is relatively compact in H^1. In particular, if $\Omega = \mathbb{R}^N$, $a \equiv a^\infty$ there exists a minimum.

Remark 1.1   Of course, $u \in H_0^1(\Omega)$ is considered as an element of $H^1(\mathbb{R}^N)$ by the usual extension by 0. And, if $I = I^\infty$ the above result means that there exist minimizing sequences (u_n) such that $u_n(\cdot + y_n)$ is relatively compact in H^1 for some sequence $(y_n)_n$ in $\mathbb{R}^N$ satisfying: $|y_n| \to \infty$ as $n \to \infty$.

Remark 1.2   If $a(x) \le a^\infty$ on $\mathbb{R}^N$, $\underset{\mathbb{R}^N}{\text{Min}}\, a(x) < a^\infty$, then one can show easily that $I < I^\infty$. One can also check that $I < I^\infty$ if a satisfies (we consider to simplify only the case when $N \ge 3$)

$$\{a^\infty - a(x)\} \exp\{\mu\sqrt{a^\infty}\,|x|\} \to +\infty \qquad \text{as } |x| \to \infty$$

for some $\mu \in (0, 1)$. On the other hand, if $a \ge a^\infty$ and $a \not\equiv a^\infty$, then clearly $I = I^\infty$ and one shows that I does not have a minimum (and thus $|y_n|_n \to \infty$ for all minimizing sequences ...).

Remark 1.3   In fact, for more general problems, one may imagine other possible ways of losing compactness. For instance, one can build sequences of the form $u_n^1 + u_n^2$ where u_n^1, u_n^2 have compact support (say), dist (Supp u_n^1, Supp u_n^2) $\to \infty$,

$$\int_{\mathbb{R}^N} |u_n^1|^{p+1} dx \underset{n}{\to} \Theta, \quad \int_{\mathbb{R}^N} |u_n^2|^{p+1} dx \underset{n}{\to} (1 - \Theta)$$

and $\Theta \in (0, 1)$. This leads to the comparison between $I = I_1$ and $I_\Theta + I_{1-\Theta}^\infty$ where I_λ, (resp. I_λ^∞) denotes the value of the same infima as in (3) (resp. (8)) but with 1 replaced by λ as a constraint. This is why, in general problems, condition (9) is replaced by $I_1 < I_\Theta + I_{1-\Theta}^\infty$. Here, since we have clearly

$I_\lambda = \lambda^{2/(1+p)} I_1$, $I_\lambda^\infty = \lambda^{2/(1+p)} I_1^\infty$, only (9) occurs.

We will not prove Proposition 1.1 here: let us only mention that its proof is a direct consequence of a general lemma on arbitrary sequences of probability measures on $\mathbb{R}^N$ (say!). We apply, in the course of proving Proposition 1.1, the lemma below with $P_n = |u_n|^{p+1}$, where u_n is a minimizing sequence of (3).

<u>Lemma 1.1</u>   Let $(P_n)_n$ be a sequence of probability measures on $\mathbb{R}^N$. Then, there exists a subsequence $(P_{n_k})_k$ satisfying one of the following three possibilities:

(i)   (compactness) P_{n_k} is tight up to a translation i.e. there exists $(y_k)_k$ in $\mathbb{R}^N$ such that

$$\forall \varepsilon > 0, \ \exists R < \infty, \ P_{n_k}(B(y_k, R)) > 1 - \varepsilon \ ;$$

(ii)   (vanishing) $\forall R < \infty, \ \underset{y \in \mathbb{R}^N}{\text{Sup}} \ P_{n_k}(B(y, R)) \to 0$;

(iii)   (dichotomy) $\exists \Theta \in (0, 1), \ \forall \varepsilon > 0, \ \forall M < \infty, \ \exists R_0 \ge M, \ R_k \underset{k}{\to} \infty$, $y_k \in \mathbb{R}^N$ satisfying

$$\left| P_{n_k}(B(y_k, R_0)) - \Theta \right| \le \varepsilon, \ \left| P_{n_k}(\mathbb{R}^N - B(y_k, R_k)) - (1 - \Theta) \right| \le \varepsilon.$$

<u>Remark 1.4</u>   The awkward form of (iii) basically means that one can split (up to ε) the measure P_{n_k} into two positive measures of mass Θ, $1 - \Theta$, supported respectively in $B(y_k, R_0)$, $B(y_k, R_k)^c$, that is, two sets going infinitely away from each other.

We now turn to the situation when a, Ω <u>are invariant</u> by a group of orthogonal transformations G. We then consider (4). We denote following [6]

$$m(x) = \# \{g \cdot x / g \in G\} \quad \forall x \in \overline{\Omega}, \quad m = \lim \inf_{x \in \overline{\Omega}, |x| \to \infty} m(x)$$

and we assume (to simplify)

$$\exists R_1 < \infty, \ \mathbb{R}^N - \overline{\Omega} \subset B(0, R_1). \tag{10}$$

Proposition 1.2   We assume (2), (6), (10) and that a, Ω are invariant by G. Then, if m = +∞, any minimizing sequence of (4) is relatively compact in $H_0^1(\Omega)$. If m < ∞, the condition

$$I_G^\infty < m\, I_{1/m}^\infty = m^{(p-1)/(p+1)}\, I^\infty \tag{11}$$

is necessary and sufficient for the compactness of all minimizing sequences.

Remark 1.5   In fact, one always has $I_G \le m^{(p-1)/(p+1)}\, I^\infty$ and if the equality holds there exist non-compact minimizing sequences and any such sequence breaks into m parts going to infinity, which are images of each other by elements of G.

Remark 1.6   As in Remark 1.5, we would like to point out that in general one has to consider the inequality $I_G = I_G(1) < I_G(\Theta) + m\, I_{(1-\Theta)/m}^\infty$ where $I_G(\lambda)$ denotes the infimum (4) with 1 replaced by $\lambda > 0$. Since $I_G(\lambda) = \lambda^{2/(p+1)}\, I_G$, $I_\lambda^\infty = \lambda^{2/(p+1)}\, I^\infty$, this general condition reduces to (11).

Remark 1.7   Of course, one has m = +∞ if a, Ω are spherically symmetric or if a, Ω are spherically symmetric with respect to x_i (for $1 \le i \le p$) where $x = (x_1, \ldots, x_p)$, $x_i \in \mathbb{R}^{N_i}$ and $N_i \ge 2$. This explains the compactness used in W. Strauss [12], H. Berestycki and P. L. Lions [5], P. L. Lions [7].

   The proof of Proposition 1.2 is a combination of Lemma 1.1 and of the G-invariance of the minimizing functions. For instance, if $(u_n)_n$ is a minimizing sequence which is compact up to a translation $(y_n)_n$ and if $m \ge 2$, we claim that the G-invariance of $(u_n)_n$ implies that $(y_n)_n$ is bounded. Indeed if $|y_n|_n \to \infty$, extracting a subsequence if necessary, we may find $g_1, \ldots, g_m$ in G, distinct and such that $g_i y_n \ne g_j y_n$ if $i \ne j$. On the other hand, there exists R_0 such that

$$\int_{B(y_n, R_0)} |u_n|^{p+1} dx \ge \frac{3}{4}$$

thus by the invariance of u_n

$$\forall i, \quad \int_{B(g_i y_n, R_0)} |u_n|^{p+1} dx \geq \frac{3}{4}.$$

This is not possible since, for n large, the balls $B(g_i y_n, R_0)$ are disjoint and $\int_{\mathbb{R}^N} |u_n|^{p+1} dx = 1$.

2. THE LIMIT CASE

We now consider the case $p = \dfrac{N+2}{N-2}$, $N \geq 3$. We will use the Hilbert space $\mathcal{D}_0^{1,2}(\Omega)$ completion of $\mathcal{D}(\Omega)$ under the norm $\left| \nabla \phi \right|_{L^2(\Omega)}$: recall that $\mathcal{D}_0^{1,2}(\Omega) = \{ u \in L^{p+1}(\Omega), \nabla u \in L^2(\Omega), u = 0 \text{ on } \partial\Omega \}$. We still consider the minimization problem (3) and we assume, in addition to (2),

$$\exists \nu > 0, \ \forall \phi \in \mathcal{D}(\Omega), \ \int_\Omega |\nabla\phi|^2 + a\phi^2 \, dx \geq \nu \int_\Omega |\nabla\phi|^2 dx \qquad (12)$$

$$a^- \in L^{N/2}(\Omega). \qquad (13)$$

Notice that (2), (12) imply $a^\infty \geq 0$. Strictly speaking, (3) is not correct if $a^\infty = 0$: in this case the minimizing class is $\{ u \in \mathcal{D}_0^{1,2}(\Omega), \ a^+ u^2 \in L^1 \}$.

Exactly as in the preceding section, we have

$$I \leq \operatorname{Inf}\{ \int_{\mathbb{R}^N} |\nabla\phi|^2 + a^\infty \phi^2 \, dx / \phi \in \mathcal{D}(\mathbb{R}^N), \ \int_{\mathbb{R}^N} |\phi|^{p+1} dx = 1 \}.$$

However, now in the limit case one can also lose compactness because of dilations. This is why we have to compute for $y \in \Omega$, $\phi \in \mathcal{D}(\mathbb{R}^N)$

$$\lim_{\sigma \to 0} \ \mathcal{E}(\sigma^{-(N-2)/2} \phi((\cdot - y)/\sigma)) = \int_{\mathbb{R}^N} |\nabla\phi|^2 dx$$

while clearly $\int_{\mathbb{R}^N} |\sigma^{-(N-2)/2} \phi((\cdot - y)/\sigma)|^{p+1} dx = 1$ and

$$\operatorname{Supp}\{ \sigma^{-(N-2)/2} \phi((\cdot - y)/\sigma) \} \subset \Omega \text{ for } \sigma > 0 \text{ small enough.}$$

This shows that $I \leq I^\infty$ where

$$I^\infty = \operatorname{Inf}\{ \int_{\mathbb{R}^N} |\nabla\phi|^2 dx / \phi \in \mathcal{D}^{1,2}(\mathbb{R}^N), \ \int_{\mathbb{R}^N} |\phi|^{p+1} dx = 1 \}.$$

53

<u>Proposition 2.1</u>   We assume (2), (12), (13) and if Ω is unbounded we assume in addition (5). Any minimizing sequence $(u_n)_n$ is relatively compact in $\mathcal{D}_0^{1,2}(\Omega)$ up to a translation and a scale change, i.e., there exist $(y_n)_n$ in $\mathbb{R}^N$, $(\sigma_n)_n$ in $]0, \infty[$ such that $\sigma_n^{-(N-2)/2} u_n((\cdot-y)/\sigma_n)$ is relatively compact in $\mathcal{D}^{1,2}(\mathbb{R}^N)$. In particular, if $\Omega = \mathbb{R}^N$, $a \equiv 0$, (3) has a minimum.

If Ω is bounded, any non-compact minimizing sequence $(u_n)_n$ satisfies: $|\nabla u_n|^2$, $|u_n|^{p+1}$ converges weakly to $I^\infty \delta_y$, δ_y for some $y \in \bar{\Omega}$ and u_n converges weakly to 0.

Finally, the condition (9) is necessary and sufficient for the compactness of all minimizing sequences in $\mathcal{D}_0^{1,2}(\Omega)$.

<u>Remark 2.1</u>   The case Ω is bounded is contained in the work of H. Brezis and L. Nirenberg [4] (see also the related work of T. Aubin [1] on Yamabe's problem). The fact that (3) has a minimum if $\Omega = \mathbb{R}^N$, $a \equiv 0$, was obtained by Rodemich [11], T. Aubin [2], G. Talenti [13]. Let us also mention that (9) is discussed in detail in [4] if Ω is bounded; and that, if $a \geq 0$, (3) does not have a minimum (and $I = I^\infty$) except if $\Omega = \mathbb{R}^N$, and $a \equiv 0$.

<u>Remark 2.2</u>   If $I = I^\infty$, there exist non-compact minimizing sequences and we can define the behaviour of such a sequence $(u_n)_n$: if Ω is bounded (up to subsequences), $v_n = \sigma_n^{-(N-2)/2} u_n((\cdot-y_n)/\sigma_n)$ converges to a minimum of I^∞ (which are explicitly known) with $\sigma_n \to +\infty$, $-y_n/\sigma_n \to y \in \bar{\Omega}$. If Ω is unbounded, v_n converges also to a minimum of I^∞ for some sequences (y_n), (σ_n).

<u>Remark 2.3</u>   Exactly as in Remark 1.3, for general problems, (9) has to be replaced by a series of strict inequalities; see [8] for more details.

The main new ingredient in the proof of Proposition 2.2 (compared with the proof of Proposition 1.1) is the following:

<u>Lemma 2.1</u>   Let $(u_n)_n$ be a sequence in $\mathcal{D}^{1,2}(\mathbb{R}^N)$ converging weakly to u and assume that $|\nabla u_n|^2$, $|u_n|^{p+1}$ converge weakly (in the sense of measures) to

some nonnegative, bounded measures μ, ν. Then, there exists an at most countable set J and two families of distinct points $(x_j)_{j \in J}$, positive numbers $(\nu_j)_{j \in J}$ such that

$$\nu = |u|^{p+1} + \sum_{j \in J} \nu_j \delta_{x_j} \tag{14}$$

$$\mu \geq |\nabla u|^2 + I^\infty \sum_{j \in J} \nu_j^{\frac{N-2}{N}} \delta_{x_j} . \tag{15}$$

We finally conclude with the situation when a, Ω are invariant by a group G of orthogonal transformations and we assume (to simplify) that Ω is bounded. Exactly as before, we consider

$$m(x) = \#\{ g \cdot x / g \in G \} , \quad \forall x \in \overline{\Omega}, \quad m = \inf_{x \in \overline{\Omega}} m(x) .$$

To explain the result which follows, let us observe that the preceding result shows that, if Ω is bounded, the only way a minimizing sequence $(u_n)_n$ may "lose compactness" is by concentration at a point $y \in \overline{\Omega}$ and formation of a Dirac mass. Now, if $(u_n)_n$ is invariant by G "necessarily", $(u_n)_n$ concentrates on the orbit of y, i.e. $\{g \cdot y / g \in G\}$, and thus at least m Dirac masses appear ... .

Proposition 2.2   We assume (2), and that Ω is bounded, a, Ω are invariant by a group G of orthogonal transformations. Then, if $m = \infty$, any minimizing sequence of (4) is relatively compact in $H_0^1(\Omega)$. If $m < \infty$, the inequality (11) is necessary and sufficient for the compactness of all minimizing sequences of (4).

Remark 2.4   (11) is discussed in [8].

REFERENCES

1.   T. Aubin, Nonlinear Analysis on Manifolds. Monge-Ampère Equations, Springer, New York, 1982.

2.   T. Aubin, Problèmes isopérimétriques et espaces de Sobolev, <u>J. Diff. Geom.</u> <u>11</u> (1976) 573-598; see also <u>C. R. Acad. Sci. Paris</u> <u>280</u> (1975) 279-282.

3.   H. Brézis, Lectures at the Berkeley Conf. on Nonlinear Problems.

4.   H. Brézis and L. Nirenberg, Positive solutions of nonlinear elliptic equations involving critical Sobolev exponents, <u>Comm. Pure Appl. Math.</u> <u>36</u> (1983) 437-477.

5.   H. Berestyki and P. L. Lions, Nonlinear scalar field equations, <u>Arch. Rat. Mech. Anal.</u> <u>82</u> (1983) 313-376; see also <u>C. R. Acad. Sci. Paris</u> <u>288</u> (1979) 395-398 and <u>287</u> (1978) 503-506.

6.   C. V. Coffman and M. Markus, Existence theorems for superlinear elliptic Dirichlet problems in exterior domains. Preprint.

7.   P. L. Lions, The concentration-compactness principle in the Calculus of Variations. The locally compact case, <u>Ann. I. H. P. Anal. Non Lin.</u> <u>1</u> (1984) 109-145 and <u>1</u> (1984) 223-283; see also <u>C. R. Acad. Sci. Paris</u> <u>294</u> (1982) 261-264 and in "<u>Contributions to Nonlinear Partial Differential Equations</u>", Pitman, London, 1983.

8.   P. L. Lions, The concentration-compactness principle in the Calculus of Variations. The limit case, <u>Rev. Mat. Iberoamer.</u> <u>1</u> (1984) ; see also <u>C. R. Acad. Sci. Paris</u> <u>296</u> (1983) 634-648 and in <u>Séminaire Goulaovic-Meyer-Schwartz</u> 1982-1983, Ecole Polytechnique, Palaiseau.

9.   P. L. Lions, Minimization problems in $L^1(\mathbb{R}^N)$, <u>J. Funct. Anal.</u> <u>41</u> (1981) 236-275.

10.   P. L. Lions, Symétrie et compacité dans les espaces de Sobolev, <u>J. Funct. Anal.</u> <u>49</u> (1982) 315-334.

11.   Rodemich, unpublished manuscript.

12.   W. Strauss, Existence of solitary waves in higher dimensions, <u>Comm. Math. Phys.</u> <u>55</u> (1977) 149-162.

13.   G. Talenti, Best constant in Sobolev inequality, <u>Ann. Mat. Pura Appl.</u> <u>110</u> (1976) 353-372.

14.   K. Uhlenbeck, Variational problems for gauge fields, in <u>Seminar on Differential Geometry,</u> ed. S. T. Yan, Princeton Univ. Press, Princeton, N.J., 1982.

Pierre-Louis Lions
Ceremade
Université de Paris IX Dauphine
Place de Lattre de Tassigny
75775 Paris Cedex 16
France

R J DIPERNA
Degenerate systems of conservation laws

We shall discuss some work dealing with singularities and oscillations in solutions to systems of conservation laws in one space dimension

$$u_t + f(u)_x = 0. \tag{1}$$

Here the state variable is constrained to a region G of R^n which serves as the domain of definition of the nonlinear flux function f:

$$u = u(x, t) \in G \subset R^n.$$

The theory for systems of conservation laws (1) has developed along two related directions. The geometric theory has dealt with the structure of the solution on various length and time scales. The associated analysis takes place in the strong topology and is concerned with the three main temporal stages in the evolution. Specifically, it is concerned with the development of singularities during the initial stage of the evolution from smooth data, with the propagation and interaction of singularities during the intermediate stage of the evolution and with the large-time decay and asymptotic behaviour during the final stage. The analysis of the local structure of the solution requires observations of the solution on lower dimensional sets, specifically on sets with finite one-dimensional Hausdorff measure that support the basic singularities (shock waves). The geometric theory has greatly benefited from the technical tools of geometric measure theory in general and the theory of functions of bounded variation in particular. We refer the reader to [1-13] and to the references cited therein for results on the geometric side of the subject.

In contrast, the more recently developed functional analytic theory has dealt with averaged behaviour. The analysis takes place in the weak topology and is concerned with the development and propagation of oscillations. Here one employs

57

observations of the solution on sets with positive two-dimensional Lebesgue measure. One of the main tools is the theory of compensated compactness developed by L. Tartar and F. Murat [14, 15, 16].

A few words of background are appropriate in connection with the geometric side of the subject. The geometric theory focuses attention on two classes of systems, distinguished by the structure of the eigenvalues. The main structural hypotheses for the theory of non-degenerate systems are the following. First, there exist n distinct speeds of propagation for signals. Technically, the Jacobian of f has n real and distinct eigenvalues:

$$\nabla f(u) : \lambda_1(u) < \lambda_2(u) < \ldots < \lambda_n(u). \tag{2}$$

Second, the wave speeds are monotone functions of the wave amplitude. Technically, the eigenvalues λ_j are monotone in the corresponding eigendirection, which represent the wave front in state space:

$$r_j \nabla \lambda_j \neq 0 \quad \text{where} \quad \nabla f \, r_j = \lambda_j r_j. \tag{3}$$

Thus, the framework for non-degenerate systems concerns systems which satisfy the condition (2) of strict hyperbolicity and the condition (3) of genuine non-linearity in the sense of Lax [9]. Within the framework the geometric theory is fairly well developed for solutions with small initial data, specifically for solutions whose initial data have small total variation. A number of results are available on existence, propagation of singularities and large time asymptotic behaviour. We shall not attempt to discuss this work here but concentrate on some more recent progress dealing with solutions with large data. In particular, we shall describe the first large data results for the special systems of mechanics.

We shall first recall some relevant background dealing with the small data situation for general systems of equations. The fundamental existence theorem due to Glimm [7] constructs solutions to the Cauchy problem with data near a constant state: if the total variation of the initial data u_0 is sufficiently small then there exists a globally defined distributional solution u. The solution u is generated as the limit of a sequence of finite difference approximations $u_{\Delta x}$, the so-called random choice approximations. During the course of convergence the random choice approximations maintain a uniform bound on their amplitude as measured by the L^∞ norm and their gradient as measured by the spatial total variation norm:

$$u = \lim_{\Delta x \to 0} u_{\Delta x}$$

$$\left| u_{\Delta x}(\cdot, t) \right|_\infty \leq \text{const.} \left| u_0 \right|_\infty \tag{4}$$

$$TVu_{\Delta x}(\cdot, t) \leq \text{const. } TVu_0 . \tag{5}$$

Estimates corresponding to (4) and (5) hold in the limit for the exact solution u and establish $L^\infty \cap BV$ as the natural space for the solution operator.

The random choice method has provided the main tool for the construction and analysis of solutions. The scheme generates solutions to the Cauchy problem as a limit of random choice approximations and presents a detailed picture of wave interactions which is instrumental in the study of singularities and asymptotic behaviour. The local component of this semi-analytical method consists of the Lax solution of the Riemann problem [9], i.e. the fundamental solution, and is employed in a time-marching fashion to generate an approximate solution from piecewise constant initial data.

Certain difficulties arise in the attempt to apply the random choice method to the Cauchy problem with large data. They stem from the fact that the random choice method makes explicit use of the fundamental solution of (1), in an iterative fashion to general solutions from general data. The analysis of random choice approximations requires sharp estimates on the interaction between two fundamental solutions, specifically sharp estimates relating the magnitudes of outgoing waves in an interaction to the magnitudes of incoming waves in an interaction. The sum of the magnitudes of the waves constitutes the total variation norm. The analysis of the fundamental solution and its interactions become quite complex in the regime of large data, even for nondegenerate systems. Additional complications arise in the presence of degeneracies in the eigenvalues, a fact with which one must contend in treating the systems of mechanics. This motivates in part the study of alternative construction procedures.

Here we shall describe some new results on large data existence for the (degenerate) systems of mechanics using the method of vanishing viscosity. With regard to the structural hypotheses we note that, from the point of view of wave propagation, two main degeneracies occur in mechanics. The first is the loss of strict hyperbolicity at the boundary ∂G of the state space G. This is exemplified in gas dynamics by the compressible Euler equations,

$$\rho_t + m_x = 0$$
$$m_t + (m^2/\rho + p)_x = 0 ,$$

(6)

which describe the conservation of mass and momentum. For a typical gas, the pressure p responds to the density ρ in an increasing and convex fashion. The prototypical behaviour is given by a polytropic gas, $p = A\rho^\gamma$, $\gamma > 1$. Here the eigenvalues remain distinct for $\rho > 0$ but coalesce at the vacuum line $\rho = 0$, representing the boundary of the state space:

$$\lambda_1 = u - \sqrt{p'}$$
$$u = m/\rho = \text{velocity} .$$
$$\lambda_2 = u + \sqrt{p'}$$

Physically, the collapse of the eigenvalues manifests itself in an increased coupling between the nonlinear modes as the vacuum state is approached. Indeed, the Glimm interaction coefficients, which describe the magnitude of the coupling, approach infinity as ρ vanishes.

The second form of degeneracy is associated with the loss of monotonicity of the eigenvalues on some interior manifold of the state space. This is exemplified in dynamic elasticity by the quasilinear wave equation for the displacement w:

$$w_{tt} = \sigma(w_x)_x .$$

(7)

The second order equation (7) can be recast as a first order system by introducing the state variables of velocity, $u \equiv w_t$, and strain, $v \equiv w_x$:

$$u_t - \sigma(v)_x = 0$$

$$v_t - u_x = 0 .$$

(8)

Here the stress σ typically responds to the strain v in an increasing but non-convex fashion, switching from convex in the expansive phase $v > 0$ to concave in the compressive phase $v < 0$:

$$v\sigma''(v) > 0 \quad \text{if} \quad v \neq 0 .$$

(9)

Consequently, the eigenvalues

$$\lambda_1 = -\sqrt{\sigma'} \quad \text{and} \quad \lambda_2 = \sqrt{\sigma'}$$

are not monotone. In this connection it is perhaps fair to say that one of the principal distinctions between a fluid and a solid stems from the following fact. A fluid admits shock waves only in the compressive phase of the motion; the equation of state is convex. A solid may admit shock waves in both the compressive and expansive phases of the motion; the equation of state is non-convex. The eigenvalues are obtained by differentiation. For a fluid the wave speeds λ_j are monotone, for a solid they are not monotone in general.

Thus, two prototypes for degenerate systems are provided by the equations of gas dynamics which are genuinely nonlinear but not strictly hyperbolic and by the equations of elasticity which are strictly hyperbolic but not genuinely nonlinear. For gas dynamics the difficulties stem from the unbounded coupling between the characteristic modes in the neighbourhood of the vacuum. For elasticity the difficulty stems from the complex structure of the fundamental solution which involves compound waves; the coupling is uniformly bounded in the state space.

Before presenting two representative results from the general theory we shall comment on several open problems dealing with existence and qualitative behaviour of solutions. One problem is to establish global existence of solutions to the Cauchy problem with large data for general non-degenerate and degenerate

61

systems of equations. A second related problem is to establish a priori estimates on the solution, even in the regime of small data. The only estimates which are currently available are derived from discrete versions on the random choice approximations and passing to the limit. In the absence of a priori bounds, the theory for systems of conservation laws has not yet yielded to a treatment with the standard tools of functional analysis, which require some form of control on both the amplitude and derivatives in order to establish existence. However, a new functional analytic approach to several problems in conservation laws has recently been developed with the aid of the theory of compensated compactness. The approach requires a priori control only on the amplitude of the solution and involves a study of averaged quantities and the weak topology. A third related problem is concerned with the analysis of singular perturbations and the relationship between the microscopic and macroscopic descriptions of the classical fields. Within this general area a specific problem is to prove convergence of the viscosity method with large or small data for general systems of equations.

Next we shall discuss two results dealing with large data existence using the viscosity method on degenerate systems in mechanics. Consider either the compressible Euler equations (6) with a polytropic gas in the physical range $\gamma = 1 + 2/m$ or dynamic elasticity (8) with a hard spring equation of state satisfying (9). Apply the method of artificial viscosity

$$u_t + f(u)_x = \varepsilon u_{xx} ,$$

i. e. complete parabolic regularization in which each of the primitive variables is diffused at an equal rate. For the Cauchy problem with arbitrary initial data in L^∞ one can make the following two assertions. First, the family of flows u^ε is bounded uniformly in x, t and ε

$$\left| u^\varepsilon(x, t) \right| \le M ,$$

by a constant M depending only on the L^∞ norm of the data. This result is due to a simple maximum principle which is present at the hyperbolic level and which is preserved by precisely those diffusion operators which act on each variable at an equal rate. Second, using only this a priori control on the amplitude of the solution sequence one can extract a subsequence u^{ε_k} which converges in the strong L^1_{loc} topology to a solution u of the associated hyperbolic system.

The proof employs four general tools: the representing measure of L. C. Young, which was first introduced into p. d. e. by L. Tartar, the theory of compensated compactness developed by L. Tartar and F. Murat, the theory of generalized entropy in the sense of Lax and the theory of geometrical optics in the form of the asymptotic analysis of high frequency solutions to linear hyperbolic p. d. e. We refer the reader to $[4, 17]$ for further discussion and details of the proof.

REFERENCES

1. Dafermos, C. M. , Characteristics in hyperbolic conservation laws. A study of the structure and asymptotic behaviour of solutions, in Nonlinear Analysis and Mechanics: Heriot-Watt Symposium Vol. 1, ed. R. J. Knops, Pitman (1977).

2. Dafermos, C. M. , Hyperbolic systems of conservation laws, in Systems of Nonlinear Partial Differential Equations, ed. J. M. Ball, D. Reidel (1983).

3. DiPerna, R. J. , Decay and asymptotic behaviour of solutions to nonlinear hyperbolic systems of conservation laws (to appear).

4. DiPerna, R. J. , Singularities of solutions of nonlinear hyperbolic systems of conservation laws, Arch. Rat. Mech. Anal. 60 (1975) 75-100.

5. DiPerna, R. J. , Convergence of approximate solutions to conservation laws, Arch. Rat. Mech. Anal. 82 (1983) 27-70.

6. DiPerna, R. J. , Uniqueness of solutions to hyperbolic conservation laws, Indiana Univ. Math. J. 24 (1975) 1047-1071.

7. Glimm, J. , Solutions in the large for nonlinear hyperbolic systems of equations, Comm. Pure Appl. Math. 18 (1965) 697-715.

8. Glimm, J. and P. D. Lax, Decay of solutions of systems of nonlinear hyperbolic conservation laws, Mem. Amer. Math. Soc. 101 (1970).

9. Lax, P. D. , Hyperbolic systems of conservation laws, II, Comm. Pure Appl. Math. 10 (1957) 537-566.

10. Lax, P. D. , Hyperbolic Systems of Conservation Laws and the Mathematical Theory of Shock Waves, CMBS, Monograph No.11, SIAM (1973).

11.   Liu, T.-P., The deterministic version of the Glimm scheme, <u>Comm.</u> <u>Math. Phys.</u>, <u>57</u> (1977) 135-148.

12.   Liu, T.-P., Admissible solutions to systems of conservation laws, <u>Amer. Math. Soc. Memoirs</u> (1982).

13.   Majda, A., Compressible fluid flow and systems of conservation laws in several space variables, preprint, Center for Pure and Applied Mathematics, University of California, Berkeley (PAM-144).

14.   Murat, F., Compacité par compensation, <u>Ann. Scuola, Norm. Sup. Pisa</u> <u>Sci. Fis. Mat.</u> <u>5</u> (1978) 69-102.

15.   Tartar, L., The compensated compactness method applied to systems of conservation laws, in <u>Systems of Nonlinear Partial Differential Equations</u>, ed. J. M. Ball, D. Riedel (1983).

16.   Tartar, L., Compensated compactness and applications to partial differential equations, in Research Notes in Mathematics No. 39, <u>Nonlinear Analysis and Mechanics: Heriot-Watt Symposium</u>, Vol. IV, ed. R. J. Knops, Pitman (1979).

17.   DiPerna, R. J., Convergence of the viscosity method for isentropic gas dynamics, <u>Comm. Math. Physics</u>, <u>91</u> (1983) 1-30.

Ronald J. DiPerna
Department of Mathematics
Duke University
Durham
North Carolina 27706
U. S. A.

S KLAINERMAN
Long time behaviour of solutions to nonlinear wave equations

Most basic equations of both physics and geometry have the form of nonlinear second-order autonomous systems

$$G(u, u', u'') = 0 \qquad (1)$$

where $u = u(x^1, x^2, \ldots, x^{n+1})$, and u', u'' denote all the first and second partial derivatives of u. For simplicity we will assume here that both u and G are scalars and denote by u_a, u_{ab} the partial derivatives $\partial_a u$ and respectively $\partial^2_{ab} u$; $a, b = 1, 2, \ldots, n+1$. Let $u^0(x)$ be a given solution of (1). Our equation is said to be elliptic or hyperbolic at $u^0(x)$ according to whether the $(n+1) \times (n+1)$ matrix, whose entries are $G_{ab} = \dfrac{\partial G}{\partial u_{ab}}(u^0, u^{0\prime}, u^{0\prime\prime})$, is nondegenerate and has signature $(1, \ldots 1, 1)$ or $(1, \ldots 1, -1)$.

Nonlinear elliptic equations and systems have received a lot of attention in the past forty or fifty years and in this period a lot of progress has been made and powerful methods have been developed. By comparison, the field of nonlinear hyperbolic equations is wide open. In what follows I will try to point out some recent developments concerning long-time behaviour of smooth solutions to a large class of such equations.

Let us assume that $G(0, 0, 0) = 0$ and that (1) is hyperbolic around the trivial solution $u^0 \equiv 0$. Typically, the operator obtained by linearizing (1) around $u^0 \equiv 0$ contains only second derivatives. Without further loss of generality we may assume it to be the wave operator $\partial^2_1 + \ldots + \partial^2_n - \partial^2_t = -\Box$, where we have ascribed to x_{n+1} the role of the time variable t. The equation (1) takes the form (1')

$$\Box u = F(u, u', u'') \qquad (1')$$

with F a smooth function of (u, u', u''), independent of u_{tt}, vanishing together with all its first derivatives at $(0, 0, 0)$.

Associate to $(1')$ the pure initial value problem

$$u(x, 0) = \varepsilon f(x), \quad u_t(x, 0) = \varepsilon g(x) \tag{1'a}$$

where f, g are C^∞-functions, decaying sufficiently fast at infinite (for simplicity, say f, g $\in C_0^\infty(R^n)$) and ε is a parameter which measures the amplitude of the data. Given f, g and F we define the life span $T_\star = T_\star(\varepsilon)$ as the supremum over all $T \geq 0$ such that a C^∞-solution of $(1')$, $(1'a)$ exists for all $x \in R^n$, $0 \leq t < T$. The fundamental <u>local existence theorem</u> ([1], [2], [3], [4], [5]), asserts that, if ε is sufficiently small, so that the initial data lies in a neighbourhood of hyperbolicity of the zero solution, then $T_\star(\varepsilon) > 0$. Moreover, a simple analysis of the proof shows that $T_\star(\varepsilon) \geq A\dfrac{1}{\varepsilon}$ where A is some small constant, depending only on a finite number of derivatives of F, f, g. This lower bound for $T_\star$ is in general sharp if the number of <u>space dimensions</u> n is equal to one. Indeed, let our variables in $(1')$ be x and t and $F = \sigma(u_x)u_{xx}$ with σ a smooth function, $\sigma(0) = 0$. An old result of P. Lax [6], extended to systems of wave equations by F. John [7], shows that, under the assumption of "genuine nonlinearity", $\sigma'(0) \neq 0$, all solutions to the corresponding initial value problem $(1'a)$ blow up by the time $0(\dfrac{1}{\varepsilon})$.

Recently, in [8], it was proved that $T_\star < \infty$ even if the genuine nonlinearity condition is violated. More precisely, assume that $\sigma'(0) = \ldots = \sigma^{(P)}(0) = 0$, $\sigma^{(P+1)}(0) \neq 0$, then the corresponding solutions blow up by the time $T = 0(\dfrac{1}{\varepsilon^{P+1}})$. In both situations the blow-up occurs in the second derivatives of u, i.e., u_{xx} becomes infinite while u_t, u_x remain bounded. Such blow-ups are typical of shock formations and are observable phenomena of physical reality. If the original equation, or system, can be written in conservation form, i.e., in our case, $F(u, u', u'') = \sum_{a=1}^{n+1} \partial_a f^a(u, u')$, one can try to extend the solutions past these breakdown points by introducing the concept of weak solutions. This was successfully accomplished for very general first-order systems of conservation laws, in one space dimension, by the fundamental work of Oleinik, P. Lax and J. Glimm (see [9] for a bibliography). In this lecture I will restrict myself, however, to classical solutions, i.e., C^{∞}-solutions.

Surprisingly, the situation looks better in higher dimensions. In 1976 F. John [10] proved that, under the assumption $F = F(u', u'')$ and $n \geq 3$, $T_{\star}(\varepsilon)$ can be significantly improved and, in 1980, S. Klainerman [11] was able to push $T_{\star}(\varepsilon)$ to infinity, and thus obtain global solutions, provided that $n \geq 6$. More generally (see [12], [13], [14]) :

Theorem 1   Assume that $F = F(u', u'') = 0(|u'| + |u''|)^{P+1}$ for small u', u'' and that $\frac{1}{P}(1 + \frac{1}{P}) < \frac{n-1}{2}$, then, there exists an ε_0 sufficiently small such that, for all $\varepsilon \leq \varepsilon_0$, (1'), (1'a) has a unique smooth solution for all $x \in R^n$, $t \geq 0$.

The reason for this improved behaviour of solutions of (1') in higher dimensions was beautifully illustrated by F. John [10] with the following quotation from Shakespeare, Henry VI:

"Glory is like a circle in the water,

Which never ceaseth to enlarge itself,

Till by broad spreading it disperse to nought".

Indeed, the higher the dimension, the more room for waves to disperse and thus decay. Accordingly, the key to [10]-[14] is to use decay estimates for solutions to the classical wave equation, $\Box u = 0$ (see [15], [16], [17]) , and to combine them with energy estimates for higher derivatives of solutions to the original nonlinear equations.

The dimension $n = 3$, to which nature gives preference, is not only the most important but also the most challenging. In [18], F. John exhibited an example for which $T_{\star} < \infty$. More precisely, consider $F = u_t \cdot u_{tt}$ and the corresponding equation (1') in three space dimensions. Imposing only one mild restriction on the data, $\int g(x)\, dx \geq 0$, F. John showed that there is no C^2-solution defined for all $x \in R^3$, $t \geq 0$. However, for sufficiently small ε, the solutions remain smooth for an extremely long time before a breakdown occurs. We have, in fact, the following very general:

<u>Theorem 2</u> (F. John - S. Klainerman). Assume that F verifies one of the following hypotheses:

(H₁) F does not depend explicitly on u, i.e., $F = F(u', u'')$.

(H₂) F can be written in conservation form,

$$F(u, u', u'') = \sum_{a=1}^{4} \partial_a f^a(u, u') .$$

(H'₂) $F(u, u', u'') = \sum_{a=1}^{4} \partial_a f^a(u, u') + 0(|u| + |u'| + |u''|)^3$, for small

u, u', u''.

Then, there exist three small positive constants, ε_0, A, B, depending only on a finite number of derivatives of F, f, g, such that, for every $0 \leq \varepsilon < \varepsilon_0$,
$T_\star(\varepsilon) \geq A\exp(B\frac{1}{\varepsilon})$.

Previously a weaker, polynomially long-time existence result was proved by F. John in [20] using an asymptotic expansion in powers of ε for u (see also [10]). The exponential long-time existence result was first proved for spherically symmetric solutions (in the semilinear case $F = F(u')$ by F. John [18] and T. Sideris [21], and for $F = F(u', u'')$ by S. Klainerman [22]).

The result of Theorem 2 is in general sharp. Indeed, F. John [23] proved recently that this is the case in the context of his previous example, $F = u_t u_{tt}$. There is, however, quite a rich class of nonlinearities F for which global existence holds. The following can be regarded as a generalization of Theorem 1 in dimension n = 3:

<u>Theorem 3</u>   Assume that F verifies the following "<u>Null</u>"-condition

(i) $\quad \sum_{a,b=1}^{4} F''_{u_a u_b}(u, u', u'') X^a X^b = 0(|u| + |u'| + |u''|)^3$ $\qquad$ (N)

(ii) $\quad \sum_{a,b,c=1}^{4} F''_{u_a u_{bc}}(u, u', u'') X^a X^b X^c = 0(|u| + |u'| + |u''|)^3$

(iii) $\quad \sum_{a,b,c,d=1}^{4} F''_{u_{ab} u_{cd}}(u, u', u'') X^a X^b X^c X^d = 0(|u| + |u'| + |u''|)^3$

68

for every sufficiently small u, u', u'' and any fixed null space-time vector (X^1, X^2, X^3, X^4), i.e., $(X^1)^2 + (X^2)^2 + (X^3)^2 - (X^4)^2 = 0$.

Then, if ε is sufficiently small, a global smooth solution of (1'), (1'a) exists.

To illustrate the content of Theorem 2, note that either of the following examples verifie: the condition (N):

Example 1 $F = u_a u_{bc} - u_b u_{ac}$, for any three indices a, b, c = 1, 2, 3, 4.

Example 2 $F = \partial_a (u_t^2 - u_1^2 - u_2^2 - u_3^2)$, for any index a = 1, 2, 3, 4.

The proof of both Theorems 2 and 3 depends on some recent [25] weighted L^∞ and L^1 estimates for solutions to the classical inhomogeneous wave equation in dimension n = 3. They were first used, in the spherical symmetric case, in [22] and then extended to the general case by introducing the angular momentum operators

$$\Omega_1 = x_3 \partial_2 - x_2 \partial_3, \quad \Omega_2 = x_1 \partial_3 - x_3 \partial_1, \quad \Omega_3 = x_3 \partial_1 - x_1 \partial_3 .$$

Their key property is that they commute with the wave operator $\square$ and thus can be treated as the usual partial derivatives ∂_1, ∂_2, ∂_3. In particular, this allows us to extend the energy estimates used in [10]-[14] to any combination of the derivatives ∂_1, ∂_2, ∂_3, Ω_1, Ω_2, Ω_3. The operators Ω_1, Ω_2, Ω_3 are closely connected to the "radiation operators" L_1, L_2, L_3 which played a fundamental role in [20].

A different and very interesting proof of Theorem 3, based on some conformal mapping methods, was given by D. Christodoulou [26] (see also his previous joint work with Y. Choquet-Bruhat [27]).

Both Theorems 2 and 3 have straightforward extensions to systems, in particular to those of the type arising in nonlinear elasticity and general relativity. There are important problems, like that of stability of the Minkowski-space as a solution of the Einstein equations in vacuum, for which we hope that Theorems 2 and 3 may be relevant. In the scalar case we believe that the picture provided by these theorems, together with the nonexistence results of F. John [18], [23], can

be completed. In other words, we conjecture that if one of the hypotheses (H1),
(H2), (H2') holds and (N) fails, then the lower bound on $T_\star(\varepsilon)$ given by
Theorem 2 is sharp, i.e., singularities must develop by that time, for any choice
of f or g and ε small.

An important open question is to describe the type of blow-up which occurs
in this case. If F is quasilinear and verifies H1, we expect that, as for $n = 1$,
the breakdown occurs when the second derivatives of u become infinite while
the first derivatives remain bounded. The recent work of **F**. John [28] points in
this direction, but completely satisfactory results are still missing. Another
open question is to derive results similar to Theorems 2 and 3 for the dimensions
$n = 2$ and $n = 4$. We suspect that the corresponding optimal lower bound for $T_\star$
for $n = 2$ must be $0(\frac{1}{\varepsilon^2})$, while for $n = 4$ one should be able to prove global
existence. In this respect we hope to find decay estimates similar to those of
[25] for $n \neq 3$. The same types of question can be asked for equations (1') where
the wave operator $\square$ is replaced by the Klein-Gordon operator, $\square + m^2$, or the
Schrödinger operator $-i\partial_t + \Delta$. General results of the type of Theorem 1 were
derived in [12], [13], [14], and for nonlinearities depending only on u in [29]
(see also the references there). The methods used to derive Theorems 2 and 3
might be used to improve these results substantially.

Finally, I must apologize for not mentioning the work of many other authors.
In particular, I have left out a lot of interesting results concerning semilinear
equations, i.e. $F = F(u)$ in (1'). For an up-to-date bibliography concerning
such results I refer to the recent papers of R. Glassey [30], [31].

REFERENCES

1.   Courant, R., Friederichs, K. O., and Lewy, H. Uber die partiellen
        Differentialgleichungen der mathematischen Physik, Math. Annalen, 100
        (1928) 32-74.

2.   Schauder, J. Das Anfangswertproblem einer quasilinearen hyperbolischen
        Differentialgleichung zweiter Ordnung in beliebiger Anzahl von
        unabhangingen Veranderlichen, Fundamenta Mathematica, 24 (1935)
        213-246.

3.   Friederichs, K. O. Symmetric hyperbolic linear differential equations,
        Comm. Pure Appl. Math., 7 (1954) 345-392.

4. Laray, J. Hyperbolic Differential Equations, Princeton (1952) I. A. S. ed.

5. Kato, T. Linear and quasilinear equations of evolution of hyperbolic part, C. I. M. E. II CICLO (1976).

6. Lax, P. D. Development of singularities of solutions of nonlinear hyperbolic partial differential equations, J. Math. Phys. 5 (1964) 611-613.

7. John, F. Formation of singularities in one-dimensional nonlinear wave propagation, Comm. Pure Appl. Math. 27 (1974) 377-405.

8. Klainerman, S. and Majda, A. Formation of singularities for wave equations including the nonlinear vibrating string, Comm. Pure Appl. Math. 33 (1980) 241-263.

9. Lax, P. Hyperbolic Systems of Conservation Laws and the Mathematical Theory of Shock Waves, C. B. M. S. Monograph No. 11, SIAM (1973).

10. John, F. Delayed singularity formation in solutions of nonlinear wave equations in higher dimensions, Comm. Pure Appl. Math. , 29 (1976) 649-681.

11. Klainerman, S. Global existence for nonlinear wave equations, Comm. Pure Appl. Math. , 33 (1980) 43-101.

12. Klainerman, S. Long-time behavior of solutions to nonlinear evolution equations, Arch. Rat. Mech. and Anal. , 78 (1982) 73-98.

13. Shatah, J. Global existence of small solutions to nonlinear evolution equations, Preprint.

14. Klainerman, S. , and Ponce, G. Global small amplitude solutions to nonlinear evolution equations, Comm. Pure Applied Math. , 36 (1983) 133-141.

15. Von Wahl, W. L^P-decay rates for homogeneous wave equations, Math. A. , 120 (1971) 93-106.

16. Marshall, B. , Strauss, W. and Wainger, S. L^P-L^q estimates for the Klein-Gordon equation, J. Math. Pure Appl. , 59 (1980) 417-440.

17. Pecher, H. L^P-Abschatzungen und klassiche Lösungen für nichtlineare Wellengleichungen, Math. Z. , 150 (1976) 159-183.

18. John, F. Blow-up for quasilinear wave equations in three space dimensions, Comm. Pure Appl. Math. , 34 (1981) 29-51.

19. John, F. and Klainerman, S. Almost global existence to nonlinear wave equations in three space dimensions, Comm. Pure Appl. Math. 37 (1984) 443-455.

20. John, F. Lower bounds for the life-span of solutions of nonlinear wave equations in three dimensions, Comm. Pure Appl. Math. , 36 (1983) 1-35.

21. Sideris, T. Global behavior of solutions to nonlinear wave equations in three space dimensions. Preprint (1982).

22. Klainerman, S. On "almost global" solutions to quasilinear wave equations in three space dimensions, Comm. Pure Appl. Math. , 36 (1983) 325-344.

23. John, F. Improved estimates for blow-up for solutions of strictly hyperbolic equations in three space dimensions (in preparation).

24. Klainerman, S. Global existence of nonlinear wave equations in three space dimensions (in preparation).

25. Klainerman, S. Weighted L^∞ and L^1 estimates for solutions to the classical wave equation in three space dimensions, Comm. Pure Appl. Math. 37 (1984) 269-288.

26. Christodoulou, D. Global existence of nonlinear equations (in preparation).

27. Choquet-Bruhat, Y. and Christodoulou, D. Existence of global solutions of the Yang-Mills, Higgs fields in 4-dimensional Minkowski space, I, II, Comm. Math. Physics, 83 (1982) 171-191, 193-212.

28. John, F. Blow-up of radial solutions of $\Box u = \partial F(u_t) / \partial u_t$ (in preparation).

29. Strauss, W. Nonlinear scattering theory of low energy, J. Funct. Anal. 41 (1981) 110-133.

30. Glassey, R. Finite time blow-up for solutions of nonlinear wave equations, Math. Z. 177 (1981) 323-340.

31. Glassey, R. Existence in the large for $\Box u - F(u)$ in two space dimensions, Math. Z. 178 (1981) 233-261.

Sergiu Klainerman
Courant Institute of Mathematical Sciences
251 Mercer Street
New York
New York 10012
U. S. A.

A AROSIO

Scattering theory for temporally inhomogeneous evolution equations in Banach space

We review the results of [4] (cf. [2]) on scattering theory for the evolution equation in <u>reflexive</u> Banach space X

$$y' + P(t) y = 0 \qquad \text{(for a. a. } t \in \mathbb{R}).\qquad (1)$$

1. GENERAL THEORY

For a. a. $t \in \mathbb{R}$, $P(t)$ is a densely defined closed linear operator in X. We make the following

<u>Assumption 1</u> $D(P(t)) = \text{const.} \stackrel{\text{def}}{=} Y$ (then Y is endowed with any graph norm $| \ |_Y \ (\geq | \ |_X)$, which renders Y a Banach space and makes $P(t)$ belong to $B(Y, X)$ for a. a. $t \in \mathbb{R}$). There exist constants λ, Λ s. t. $R(P(t) + \lambda) = X$ with $\|(P(t) + \lambda)^{-1}\|_{X, Y} \leq \Lambda$ for a. a. $t \in \mathbb{R}$.

For the general theory of Kato [7] (cf. [3]), the Cauchy problem for (1) is well-posed both in X and in Y, uniformly in $\mathbb{R}$, provided that

<u>Assumption 2</u> for a. a. $t \in \mathbb{R}$, $P(t)$ generates a C^0-group $\{e^{-sP(t)} : s \in \mathbb{R}\}$ in X; there exist $M \geq 1$ and β in $L^1(\mathbb{R})$ such that, for each $n \in \mathbb{N}$, for a. a. $t_1 < \dots < t_n$ and for each n-tuple (s_j) of positive real numbers

$$\|e^{-s_n P(t_n)} \circ \dots \circ e^{-s_1 P(t_1)}\|_X \leq M \exp(\sum_{j=1}^{n} s_j \beta(t_j)),$$

$$\|e^{s_1 P(t_1)} \circ \dots \circ e^{s_n P(t_n)}\|_X \leq M \exp(\sum_{j=1}^{n} s_j \beta(t_j)).$$

<u>Assumption 3</u> $P \in BV(\mathbb{R}, B(Y, X))$, (BV = bounded variation).

Assumption 2 could be called (X) quasi-stability on $\mathbb{R}$ in the forward as well as in the backward t-direction. As for the <u>perturbed equation,</u>

$$z' + (P(t) + \varepsilon(t))z = 0 \qquad \text{(for a.a. } t \in \mathbb{R}\text{):} \qquad\qquad (1')$$

<u>Assumption 4</u> The map $(P + \varepsilon)$ satisfies the same assumption as P; moreover

$$\int_{-\infty}^{+\infty} \|\varepsilon(t)\|_{Y,X} dt < \infty.$$

It is easy to prove $\begin{bmatrix} 6, & 8 \end{bmatrix}$ that, for every y in $L^\infty(\mathbb{R}, X)$ solution of (1), there exist (unique) $z_\pm$ in $L^\infty(\mathbb{R}, X)$ solutions of $(1')$ such that $\lim_{t\to\pm\infty} |y(t) - z_\pm(t)|_X = 0$ and vice-versa (i.e., the statement still holds true when (1) and $(1')$ are interchanged). Actually, our goal is: for every y in $L^\infty(\mathbb{R}, Y)$ solution of (1) there exist (unique) $z_\pm$ in $L^\infty(\mathbb{R}, Y)$ solutions of $(1')$ such that $\lim_{t\to\pm\infty} |y(t) - z_\pm(t)|_Y = 0$ (and vice-versa).

<u>Theorem 1</u> $\begin{bmatrix} 4 \end{bmatrix}$. Let assumptions 1–4 hold: then (2) holds true provided that at least one of the following conditions is fulfilled: (i) $P' \in BV(\mathbb{R}, B(Y, X))$; (ii) the map $\hat{P}(t) \overset{\text{def}}{=} \int_t^{t+1} P(s)\, ds$ satisfies assumption 2.

Condition (ii) amounts to saying that those general properties of P that ensure the well-posedness of the Cauchy problem must be stable with respect to the average $\int_t^{t+1} ds$. This is the case in the Hamilton equation (4) below.

2. ABSTRACT HAMILTON EQUATIONS

H, V are Hilbert spaces, $V \subset H$ densely and continuously. Identifying H with its antidual $H^\star$, we have $V \subset H \subset V^\star$, i.e. $\langle h, v \rangle = (h, v)_H$ for each $h \in H$, $v \in V (\langle\,,\,\rangle = \langle V^\star, V \rangle)$. For a.a. $t \in \mathbb{R}$, $\mathcal{H}(x, t)$ is a quadratic form on V which is positively definite, uniformly in t, i.e.

$$\mathcal{H}(x, t) = \tfrac{1}{2}\langle A(t)x, x \rangle \qquad (x \in V, \text{ for a.a. } t \in \mathbb{R}),$$

where $A(t)$ in $B(V, V^\star)$ satisfies for a.a. $t \in \mathbb{R}$ and for each $x, y \in V$

$$\langle A(t)x, y \rangle = \overline{\langle A(t)y, x \rangle}, \qquad \langle A(t)x, x \rangle \geq \nu|x|_V^2 \qquad (\nu > 0). \qquad (3)$$

J is an isomorphism of H onto itself such that $-J = J^{-1} = J^\star$ (= adjoint of

74

J in H).

Derguzov and Jakubovič [5] introduced the <u>Hamilton equation</u>

$$(Jx)' = \nabla_x \mathcal{H} = A(t)x \qquad (\text{for a. a. } t \in \mathbb{R}). \qquad (4)$$

<u>Example 1</u>   The abstract <u>wave equation</u> $u'' + A(t)u = 0$ may be written as a Hamilton equation in $\tilde{H} = H \times H$ for $x = (u, u')$, $\tilde{V} = V \times H$, $\tilde{A}(t) = \begin{pmatrix} A(t) & 0 \\ 0 & \mathbb{1}_H \end{pmatrix}$, $J = \begin{pmatrix} 0 & -\mathbb{1}_H \\ \mathbb{1}_H & 0 \end{pmatrix}$.

<u>Example 2</u>   For $J = i\mathbb{1}_H$, (4) is an abstract <u>Schrödinger equation.</u>

<u>Theorem 2</u> [4, 5].   Assume that $A \in BV(\mathbb{R}, B(V, V^\star))$. Then the Cauchy problem for (4) is well-posed in V, uniformly on $\mathbb{R}$. More precisely, for each $x_0 \in V$ there exists exactly one solution x in $C^0(\mathbb{R}, V) \cap L^\infty(\mathbb{R}, V)$ of (4) such that $x(0) = x_0$.

<u>Theorem 3</u> [4].   Assume that A, $\varepsilon \in BV(\mathbb{R}, B(V, V^\star))$, that $(A + \varepsilon)$ satisfies (3) and that $\int_{-\infty}^{+\infty} \|\varepsilon(t)\|_{V, V^\star} dt < \infty$. Then for each x in $C^0(\mathbb{R}, V)$ solution of (4) there exists (unique) $w_\pm$ in $C^0(\mathbb{R}, V) \cap L^\infty(\mathbb{R}, V)$ solutions of $(Jw)' = (A(t) + \varepsilon(t))w$ such that $\lim_{t \to \pm\infty} |x(t) - w_\pm(t)|_V = 0$ (and vice-versa).

3.   APPLICATIONS TO PARTIAL DIFFERENTIAL EQUATIONS

Let Ω be any open subset of $\mathbb{R}^n$, $(m + (\text{diam } \Omega)^{-2}) > 0$, and let the functions $\varepsilon_{ij}(x, t)$ $(i, j = 1, \ldots, n)$ satisfy $\varepsilon_{ij} = \overline{\varepsilon_{ji}}$,

$$\varepsilon_{ij} \in BV(\mathbb{R}, L^\infty(\Omega)), \quad \sum_{i, j} \varepsilon_{ij}(x, t) \xi_i \xi_j \geq (\nu - 1)|\xi|^2 \quad (\nu > 0).$$

<u>Example 3</u>   By Theorem 2 and Example 1 (cf. [1]), the Cauchy problem for

$$\Box u = mu - \sum_{ij} \frac{\partial}{\partial x_i}(\varepsilon_{ij}(x, t)\frac{\partial u}{\partial x_j}) \qquad (\text{a. e. in } \Omega \times \mathbb{R}),$$

$$u(\cdot, t) \in H_0^1(\Omega) \qquad (t \in \mathbb{R}), \qquad (5)$$

$\Box = \Delta - (\ )_{tt}$, is well-posed in $H_0^1(\Omega) \times L^2(\Omega)$, uniformly on $\mathbb{R}$. If moreover

$$\int_{-\infty}^{+\infty} \| \varepsilon_{ij}(\cdot, t) \|_{L^\infty(\Omega)} dt < \infty \qquad (i, j = 1, \ldots, n) \qquad (6)$$

then by Theorem 3 (cf. [1]) for each solution u of (5) there exist (unique)
solutions $v_\pm$ of $\{ \Box v = mv, \ v(\cdot, t) \in H_0^1(\Omega) \ \forall t \in \mathbb{R} \}$ such that

$$\lim_{t \to \pm\infty} \int_\Omega \left(|u(x, t) - v_\pm(x, t)|^2 + \sum_i \left| \frac{\partial u}{\partial x_i}(x, t) - \frac{\partial v_\pm}{\partial x_i}(x, t) \right|^2 \right) dx = 0, \quad (7)$$

$$\lim_{t \to \pm\infty} \int_\Omega \left| \frac{\partial u}{\partial t}(x, t) - \frac{\partial v_\pm}{\partial t}(x, t) \right|^2 dx = 0, \qquad (8)$$

and vice-versa (i. e. for each $v \ldots$ there exists $u_\pm \ldots$).

Example 4   By Theorem 2 and Example 2, the Cauchy problem for

$$iu_t + \Delta u = mu - \sum_{ij} \frac{\partial}{\partial x_i} \left(\varepsilon_{ij}(x, t) \frac{\partial u}{\partial x_j} \right) \quad (a.e. \text{ in } \Omega \times \mathbb{R}),$$

$$u(\cdot, t) \in H_0^1(\Omega) \qquad (t \in \mathbb{R}) \qquad (9)$$

is well-posed in $H_0^1(\Omega)$, uniformly on $\mathbb{R}$. If moreover (6) holds, then for each
solution u of (9) there exist (unique) solutions $v_\pm$ of $\{ iv_t + \Delta v = mv,$
$v(\cdot, t) \in H_0^1(\Omega) \ \forall t \in \mathbb{R} \}$ such that (7) holds, and vice-versa.

REFERENCES

1.   A. Arosio, Arch. Rat. Mech. Anal. 86 (1984), 147-180; Compt. Rend. Acad.
      Sci. Paris, Sér. I 295 (1982), 83-86.

2.   A. Arosio, Portugaliae Math. 41 (1982), 351-366.

3.   A. Arosio, Nonlinear Anal. T. M. A. 8 (1984), 997-1009.

4.   A. Arosio, Scattering theory for temporally inhonogeneous Hamilton
      equation in Hilbert space (to appear).

5.   V. I. Derguzov and V. A. Jakubovič, Soviet Math. Dokl. 4 (1963),
      1169-1172.

6.   A. Inoue, J. Math. Soc. Japan 26 (1974), No. 4, 608-624.

7.   T. Kato, J. Math. Soc. Japan 25 (1973), 648-666.

8.   E. J. P. G. Schmidt, Indiana Univ. J. 24 (1975), 925-935.

Alberto Arosio
Dipartimento di Matematica
Università
Via Buonarroti, 2
I 56100 Pisa
Italy

M BIROLI
Wiener obstacles: the parabolic case

1. NOTATION

In the following, Ω will be an open bounded set in R^N, $N \geq 3$, and $Q = \Omega \times (0, T)$. Let E be a compact set contained in a parabolic cylinder C and indicate

$$\text{cap}_C(E) = \inf \{ \int_C |D_x w|^2 \, dx \, dt \; ; \; w \in D(C), \; w \geq 1 \text{ in a}$$
$$\text{neighbourhood of } E \}$$

We have thus defined a Choquet capacity [6] and for arbitrary sets (with closure in C) we denote by cap_C the external capacity.

We observe that the following property can be proved: Let $C = \omega \times (T_0, T_1)$ and E be a capacitable set in C; we have

$$\text{cap}_C(E) = \int_{T_0}^{T_1} \text{cap}_\omega(E_t) \, dt$$

where cap_ω is the usual elliptic capacity and E_t is the section of E at the instant t.

Fixing now $z_0 = (x_0, t_0)$, we indicate

$$Q(r, z_0) = \{ (x, t) ; \; |x-x_0| < r, \; |t-t_0| < r^2 \}$$
$$Q_\theta^-(r, z_0) = \{ (x, t) ; \; |x-x_0| < (1-6\theta)^{\frac{1}{2}} r, \; t_0 - r^2 < t < t_0 - 6\theta r^2 \}$$

Let now $\psi : Q \to R \cup \{-\infty\}$ be a Borel function quasi-everywhere defined, we denote

$$E(\psi, z_0, \eta, r) = \{ z \in Q_\theta^-(r, z_0) ; \psi(z) \geq \sup \{ \psi(z) - \eta ;$$
$$|x-x_0| < r/2, \; |t-t_0| < r^2/4 \}\} \qquad \eta > 0$$
$$\Delta_\theta(\psi, z_0, \eta, r) = \Delta_\theta(\eta, r) = \text{cap}_{Q(2r, z_0)} E(\psi, z_0, \eta, r)$$
$$\delta_\theta(\eta, r) = \Delta_\theta(\eta, r) / \sigma_N r^N$$

where $\sigma_N = \text{cap}_{Q(2, z_0)} Q(1, z_0)$.

The Wiener modulus of ψ is defined by

$$\omega(r, R) = \inf\{\omega \geq 0;\ \omega \exp \int_r^R \delta_\theta(\omega, \rho)\, d\rho/\rho \geq 1\}$$

We denote, finally, by G^{z_0} the Green function of the heat operator.

2. RESULTS

All the results given here have been obtained jointly with U. Mosco.

Recently some attention has been given to certain parabolic variational inequalities with an obstacle, and to their relation to problems in stochastic optimal control [1]. In this paper we are interested in the continuity of the solution of parabolic obstacle problems; this has been studied for regular obstacles in [2], [3] in the linear case and in some nonlinear cases, and in [12] in the case of obstacles which are Hölder continuous in time and one-sided Hölder continuous in space variables.

Let $\psi : Q \to R\{+\infty\}$ be a Borel function bounded from above, which we assume to be quasi-everywhere defined (for the above-defined notion of capacity) in Q; a given function $u \in L^2(0, T; H^1(\Omega)) \cap L^\infty(0, T; L^2(\Omega))$ is a local solution of the parabolic obstacle problem relative to the heat operator and ψ if

$$\int_0^t \{D_t v(v - u)\phi + D_x u D_x(\phi(v - u))\}\, dx\, dt + \tfrac{1}{2}\int_0^t D_t(v - u)^2\, dx\, dt \geq 0$$

$\forall \phi \in C^\infty(Q)$, $\phi|_\Sigma = 0$ $(\Sigma = \partial\Omega \times (0, T))$, $\phi(x, 0) = 0$, $\phi \geq 0$,

$\forall v \in H^1(0, T; L^2(\Omega)) \cap L^2(0, T; H^1(\Omega))$, $v \geq \psi$ quasi-everywhere

in $\text{supp}(\phi) \cap \Omega \times (0, t)$.

$$\|(u - d)^+ \alpha\|^2_{L^2(\Omega)}(t_2) - \|(u - d)^+ \alpha\|^2_{L^2(\Omega)}(t_1) \leq$$
$$2 \int_{t_1}^{t_2} \{|D_x u||D_x \alpha|(u - d)^+ \alpha - |D_x(u - d)^+|^2 \alpha^2\, dx\, dt$$

$\forall \alpha \in C^\infty(\Omega)$, $d \geq \psi$ quasi-everywhere in $\text{supp}(\alpha) \times (t_1, t_2)$, $\alpha \geq 0$.

We obtain the following results:

Theorem 1   Let u be a local solution of the parabolic obstacle problem relative to the heat operator and ψ; then for $z_0 \in Q$ we have

$$M(r) \leq K \{ M(R) \omega_\theta (r, R)^\beta + \omega_\theta (r, R) \Lambda \; \text{osc}_{Q(R, z_0)} \psi \}$$

where $\beta \in (0, 1)$, $R \leq R_0$ (R_0 suitable), $\theta \in (0, \theta_0)$ and

$$M(r) = (\int_{t_0 - r^2}^{t_0} \int_{B(r; x_0)} |D_x u|^2 \, dx \, dt)^{\frac{1}{2}} + \text{osc}_{Q(r; z_0)} u \; .$$

Theorem 2   Let the assumptions of Theorem 1 hold, then

$$\text{osc}_{Q(r; z_0)} u \leq K (R^\gamma + \text{osc}_{Q(R; z_0)} \psi) \omega_\theta (r, R) + \omega_\theta (r, R) \Lambda \; \text{osc}_{Q(R; z_0)} \psi$$

where $\gamma \in (0, 1)$.

Such an estimate was first obtained for the elliptic case in [10], [11].

Corollary 1   If

$$\lim_{r \to 0} \omega_\theta (r, R) = 0 \tag{2.1}$$

then u is continuous at z_0; moreover, if

$$\omega_\theta (r, R) \leq K (r/R)^\nu \qquad \nu \in (0, 1) \tag{2.2}$$

then u is Hölder continuous at z_0.

Corollary 2   If ω_θ is continuous (Hölder continuous) at z_0, then u is continuous (Hölder continuous) at z_0.

Remark 1   We observe that $\omega_\theta (r, R)$ is increasing in θ then, if (2.1) or (2.2) holds for θ_0, it holds again for $\theta \in (0, \theta_0)$.

Remark 2   The condition (2.1) holds if

$$\lim_{r \to 0} \int_r^1 \delta_\theta (\eta, \rho) \, d\rho /\rho = +\infty \qquad \forall \eta > 0$$

The condition (2.2) holds if

$$\liminf_{r \to 0} \left| \lg r \right|^{-1} \int_r^1 \delta_\theta(\eta, \rho) \, d\rho / \rho \geqq K > 0 .$$

Remark 3   The results given here hold again for general linear parabolic operators and can be proved also in the nonlinear case with quadratic growth in the spatial gradient.

The methods used here to prove the result of Theorem 1 are derived from the analogous problem solved by J. Frehse, U. Mosco, [7], [8], [9], [10], [11], and from the proof of a Wiener estimate at boundary points for the solution of a parabolic equation [4], [5]; the main difficulties are the choice of a suitable notion of capacity (which in this case is different from the one used in [4], [5]) and the proof of a Poincaré inequality involving only the spatial gradient, which is given for local solution of our variational inequality.

REFERENCES

1.   Bensoussan, A., Lions, J. L. Applications des Inéquations Variationnelles en Contrôle Stochastique, Dunod (1978).

2.   Biroli, M. Hölder regularity for parabolic obstacle problem, Boll. U. M. I. 1-B, (1982), 1079-1088.

3.   Biroli, M. Existence d'une solution hölderienne pour des inéquations paraboliques non linéaires avec obstacle, C. R. A. S. Paris, 269 (1983), 7-9.

4.   Biroli, M., Mosco, U. Estimations ponctuelles sur le bord du module de continuité des solutions faibles des équations paraboliques, C. R. A. S. Paris

5.   Biroli, M., Mosco, U. Wiener estimates at boundary points for parabolic equations. Preprint Sonderforschungbereich 72, Universität Bonn, 1984.

6.   Choquet, G. Lectures on Analysis, Vol. 1, Benjamin (1969).

7.   Frehse, J., Mosco, U. Irregular obstacles and quasi-variational inequalities of the stochastic impulse control, Ann. Sc. Norm. Sup. Pisa IV, IX (1) (1982), 105-157.

8.   Frehse, J., Mosco, U. Sur la continuité ponctuelle des solutions faibles du problème d'obstacle, C. R. A. S. Paris, 295 (1982), 571-574.

9. Frehse, J., Mosco, U. Wiener obstacles, in <u>Nonlinear Partial Differential Equations and their Applications, Collège de France Seminar</u> vol. C, Pitman (1984).

10. Mosco, U. Module de Wiener pour un problème d'obstacle, <u>C.R.A.S. Paris</u> (to be published).

11. Mosco, U. Lecture at <u>Summer Institute on Nonlinear Functional Analysis and Applications</u> A.M.S., Berkeley, July 1983.

12. Struwe, M., Vivaldi, M.A. On the Hölder continuity of bounded weak solutions of quasi-linear parabolic variational inequalities, <u>Ann. Mat. Pura e Appl.</u> (to be published).

Marco Biroli
Dipartimento di Matematica
Politecnico
Via Bonardi, 9
I 20133 Milano
Italy

G BOTTARO & P OPPEZZI
An integral representation theorem for local functionals

Recently G. Buttazzo, G. Dal Maso and L. Modica studied functionals such as $J : W^{1,p}(\Omega) \times \mathcal{B}(\Omega) \to \mathbb{R}$, where Ω is an open set in $\mathbb{R}^n$ and $\mathcal{B}(\Omega)$ the class of its Borel subsets, assuming J to be local type (i.e., $u_{|A} = v_{|A}$ implies $J(u, A) = J(v, A)$ for each open $A \subset \Omega$, u, v $W^{1,p}(\Omega)$) and $J(u, .)$ to be a measure [1], [3].

This topic is mainly interesting in the Γ-convergence theory of integral functionals where the problem arises of finding when an integral representation of the Γ-limit, which in many cases is a functional of the above kind, is possible.

In a paper to appear we give the following result.

<u>Theorem</u>   Let X be a separable Banach space, $p \in [1, \infty]$ and let

$$J : W^{1,p}_{loc}(\Omega, X) \times \mathcal{B}(\Omega) \to [0, \infty]$$

satisfy:

(a) $J(u, .)$ is finitely additive if $u \in W^{1,\infty}_{loc}(\Omega, X)$ and is a regular measure when $u \in \mathcal{P} \equiv \{u : \Omega \to X, u(t) = \Sigma^n_{i=1} t_i x_i, x_i \in X_0, i = 1, \ldots n\}$, where X_0 denotes a numerable dense subset in X;

(b) $J(u, A) = J(v, A)$ if $A \subset\subset \Omega$, $\nabla u_{|A} = \nabla v_{|A}$, $u, v \in W^{1,\infty}_{loc}(\Omega, X)$ (by $A \subset\subset \Omega$ we mean that A is open and $\bar{A}$ is a compact subset of Ω);

(c) $J(u, A) \leq \lim_k \inf J(u_k, A)$ in each of the following cases:

 (c_0) A is an n-dimensional cube $\subset\subset \Omega$, $u, u_k \in W^{1,\infty}_{loc}(\Omega, X)$, $k \in \mathbb{N}$, $u_k \to u$ uniformly on A and $\sup_k |\nabla u_k|_{L^\infty(A, X)} < \infty$;

 (c_1)   if $p < \infty$, $u_k, u \in W^{1,p}_{loc}(\Omega, X)$, $k \in \mathbb{N}$, $A \subset\subset \Omega$, $u_{k|A} \to u_{|A}$ in in $W^{1,p}(A, X)$,

  or, if $p = \infty$, $u_k, u \in W^{1,\infty}_{loc}(\Omega, X)$, $A \subset\subset \Omega$, $u_k \to u$ uniformly on A and in $W^{1,\alpha}(A, X)$ for each $\alpha \in \,]1, \infty[$ and $\sup_k |\nabla u_k|_{L^\infty(A, X)} < \infty$;

(d) if $p < \infty$ there exists $\phi \in L^1_{loc}(\Omega)$, $c > 0$ such that

$$J(u, E) \leq c \int_E (\phi + |u|^p + |\nabla u|^p) \, dt, \forall (u, E) \in W^{1,p}(\Omega, X) \times \mathcal{B}(\Omega);$$

if $p = \infty$, for each $r > 0$ there exists a $g_r \in L^1_{loc}(\Omega)$ such that

$$J(u, E) \leq \int_E g_r \, dt \text{ when } (u, E) \in W^{1,\infty}_{loc}(\Omega, X) \times \mathcal{B}(\Omega), \ |u|_{W^{1,\infty}(E,X)} \leq r.$$

Then there exists a Caratheodory integrand $f : \Omega \times X^n \to [0, \infty[$ such that $f(t, .)$ is quasi-convex in the sense of Morrey† for almost every $t \in \Omega$ and $J(u, A) = \int_A f(t, \nabla u(t)) \, dt$ for all $u \in W^{1,p}_{loc}(\Omega, X)$, $A \subset\subset \Omega$. Moreover it satisfies the estimates:

$$f(t, y) \leq c(\phi(t) + (|t|^p + 1)|y|^p) \text{ for a.e. } t \in \Omega, \text{ and all } y \in X^n \text{ if } p < \infty;$$

$$f(t, y) \leq g_r(t) \text{ for a.e. } t \in \Omega, \text{ all } y \in X^n, |y| < r/2 \text{ if } p = \infty.$$

In [1] was given an analogous representation result in the case $X = \mathbb{R}$ and assuming $J(., A)$ convex (and so obtaining the convexity of $f(t, .)$).

Moreover G. Buttazzo and G. Dal Maso have recently told us that they obtained another improvement in the case $X = \mathbb{R}^n$ making no use of convexity [2].

REFERENCES

1. G. Buttazzo, G. Dal Maso. Γ-limits of integral functionals, J. Analyse Math. 37 (1980) 145-185.

2. G. Buttazzo, G. Dal Maso. Integral representation and relaxation of local functionals (to appear).

3. G. Dal Maso, L. Modica. A general theory of variational functionals, Sc. Norm. Sup. Pisa Topics in Functional Anslysis 1980-81, Pisa (1982) 149-221.

Gianfranco Bottaro and Pirro Oppezzi
Istituto Matematico, Università
Via L. B. Alberti, 4
I 16132 Genova
Italy

---

† We say that $g : X^n \to \mathbb{R}$ is quasi-convex in the sense of Morrey if, for every $y \in X^n$, $G \subset\subset \mathbb{R}^n$ and $z \in C^1_0(G, X)$, we have $|G|g(y) \leq \int_G g(y + \nabla z(t)) \, dt$.

I C DOLCETTA
Singular perturbation of an elliptic system and the optimal switching problem

Let us consider the following system of second-order partial differential equations:

$$-\varepsilon\Delta u_i + \lambda u_i - g_i \cdot Du_i + \sum_{\substack{j=1 \\ j\neq i}}^{m} \beta^\varepsilon(u_i - u_j - c_{ij}) = f_i \quad \text{in } \mathbb{R}^N$$

$$(i = 1, \ldots, m, \ \varepsilon > 0). \tag{1}$$

Here λ is a positive constant, g_i a given vector field on $\mathbb{R}^N$, f_i a given real valued function and $\beta^\varepsilon(x) = \beta(x/\varepsilon)$, where $\beta : \mathbb{R} \to \mathbb{R}$ is a smooth function such that

$$\beta(t) = 0 \ \text{ for } \ t \le 0, \ \beta(t) > 0 \ \text{ for } \ t > 0, \ \beta'' \ge 0, \ 0 \le \beta' \le 1 .$$

The constants c_{ij} are assumed to satisfy

$$c_{ij} > 0, \ i \neq j, \ c_{ii} = 0, \ c_{ij} < c_{il} + c_{lj}, \ i \neq j \neq l . \tag{2}$$

Under the assumptions

$$\left| g_i(x) - g_i(x') \right| \le L \left| x - x' \right|, \ \left| g_i(x) \right| \le L,$$

$$\left| f_i(x) - f_i(x') \right| \le D \left| x - x' \right|, \ \left| f_i(x) \right| \le D \tag{3}$$

it is known (see Evans-Friedman [3]) that (1) has a unique smooth solution $u^\varepsilon = (u_1^\varepsilon \ldots u_m^\varepsilon)$ for any $\varepsilon > 0$, provided that λ is large enough.

We are interested here in the behaviour of u^ε as $\varepsilon \to 0$. By maximum principle arguments one can prove (see [1]) that, for any $0 < \gamma < \min(\lambda L^{-1}, 1)$, there exists a constant C_γ such that

$$\left| u_i^\varepsilon(x) - u_i^\varepsilon(x') \right| \le C_\gamma \left| x - x' \right|^\gamma, \ C_\gamma \ \text{ independent of } \ \varepsilon ,$$

and that $\left|u_i^\varepsilon(x)\right| \le C$, for some constant C independent of ε also. Therefore, u^ε has a uniform subsequential limit as $\varepsilon \to 0$, say u. The following result has been proved in [1].

Theorem   Under the assumption made, $u = (u_1 \ldots u_m)$ is the unique bounded Hölder-continuous viscosity solution of the first-order system

$$\operatorname{Max}(\lambda u_i - g_i . Du_i - f_i;\ u_i - \min_{i \ne j}(u_j + c_{ij})) = 0 \text{ in } \mathbb{R}^N \quad (i = 1,\ldots,m). \quad (4)$$

Remark 1   The notion of a viscosity solution of first-order nonlinear partial differential equations was introduced by Crandall-Lions [2]. In the case of system (4) above, the requirement is that, for any $\Phi \in C^1(\mathbb{R}^N)$ and $i = 1, \ldots, m$, the inequality

$$\operatorname{Max}(\lambda u_i(x_0) - g_i(x_0) . D\Phi(x_0) - f_i(x_0);\ u_i(x_0) - \min_{i \ne j}(u_j(x_0) + c_{ij})) \le 0$$

holds at each local maximum point x_0 of $u_i - \Phi$ and the reverse inequality is fulfilled at any local minimum point of $u_i - \Phi$.

Remark 2   The vector $u = (u_1 \ldots u_m)$ can be interpreted in the framework of dynamic programming (see Fleming-Rishel [4]) as the value of the optimal switching problem. Namely,

$$u_i(\mathbf{x}) = \operatorname{Inf}\{ \sum_{k=1}^\infty \int_{t_{k-1}}^{t_k} f_{i_{k-1}}(y_x(s)) e^{-\lambda s} ds + c_{i_{k-1}, i_k} e^{-\lambda t_k} \}. \quad (5)$$

The infimum in (5) is to be taken over the class of admissible controls, that is, sequences of the type (t_k, i_k) such that $0 \le t_0 \le t_1 \le \ldots \le t_k \to +\infty$, and $i_k \in (1, \ldots, m)$ with $i_0 = i$. The vector $y_x(s)$ in (5) is the solution of the ordinary differential equation

$$\dot{y}(s) = g_{i_k}(y(s)),\quad t_k \le s \le t_{k+1},\quad y(0) = x$$

(see [1] for details).

REFERENCES

1. I. Capuzzo Dolcetta, L. C. Evans, SIAM J. Control and Optimization, 22 (1) (1984).

2. M. G. Crandall, P. L. Lions, Trans. Amer. Math. Soc., 277 (1983).

3. L. C. Evans, A. Friedman, Trans. Amer. Math. Soc., 253 (1979).

4. W. H. Fleming, R. Rishel, Deterministic and Stochastic Optimal Control, Springer (1975).

Italo Capuzzo Dolcetta
Dipartimento di Matematica
Università "La Sapienza"
Piazzale A. Moro, 5
I 00185 Roma
Italy

M CARRIERO & E PASCALI
Uniqueness for the elastic one-dimensional bounce problem

0. In [3] we studied the elastic bounce problem for a material point subject to a strength f depending on time t and for which the free motion is strained by the t-axis which forces the motion above the half-line $x \leq 0$, without friction and with perfectly elastic collisions. Now we give a new result (Theorem 2) on the uniqueness for the previous problem. This result follows from a convergence theorem for a sequence of Cauchy's problems for first-order partial differential equations (Theorem 3).

1. Let $[0, T] \subset R$ and $f \in L^1(0, T)$; a Lipschitz function $x = x(t)$ is a solution of the elastic one-dimensional bounce problem if

$$(P) \begin{cases} \text{(i) } x \leq 0 \text{ on } [0, T]; \text{ (ii) } \ddot{x} - f \leq 0 \text{ in the sense of distributions;} \\ \text{(iii) } \mathrm{supp}(\ddot{x} - f) \subseteq \{t \,|\, x(t) = 0\}; \text{ (iv) } |\dot{x}_{\pm}(t_1)|^2 - |\dot{x}_{\pm}(t_2)|^2 = \\ 2 \int_{t_1}^{t_2} f(\eta)\dot{x}(\eta)\,d\eta \quad \text{for all } t_1, t_2. \end{cases}$$

The pair (x^0, p^0) is an admissible initial condition (a.i.c.) if $x^0 < 0$, $p^0 \in R$ or $x^0 = 0$, $p^0 \leq 0$. A function $x = x(t)$ is said to be a solution of the Cauchy problem for the elastic bounce problem with (x^0, p^0) as a.i.c. if $x(t)$ satisfies (P) and $x(0) = x^0$, $\dot{x}_+(0) = p^0$.

2. We have the following existence theorem by means of non-convex penalization:

<u>Theorem 1</u> ([3]) Let $f \in L^1(0, T)$, let (x^0, p^0) be an a.i.c. and let (ψ_h) be a sequence of real functions such that:

$$\psi_h \in C^0(R); \tag{1}$$

$$\psi_h(\xi) = 0 \text{ for } \xi \leq 0; \quad \psi_h(\xi) > 0 \text{ for } \xi > 0; \tag{2}$$

$\psi_h \rightrightarrows +\infty$ on the compact subsets of $]0, +\infty[$; $\qquad$ (3)

$$\lim_{\substack{h \to +\infty \\ \xi \to 0^+}} \psi_h(\xi) / \int_0^\xi \psi_h(\eta)\, d\eta = +\infty. \qquad (4)$$

If a sequence (x_h) of solutions of the problems (P_h) $\ddot{x}(t) + \psi_h(x(t)) = f(t)$ on $[0, T]$; $x(0) = x^0$, $\dot{x}(0) = p^0$ is given, then there exists a subsequence which is uniformly convergent to a solution of the Cauchy problem for the elastic bounce problem with (x^0, p^0) as a. i. c.

3. Uniqueness for the Cauchy's problem fails also for regular $f(t)$ ($[3]$). The uniqueness problems studied are of the following kinds:
(pb1) find spaces of functions or distributions for $f(t)$ for which uniqueness holds;
(pb2) let $f(t)$ be given in some space; compute the two-dimensional Lebesgue measure for the set of a. i. c. for which the uniqueness fails.

Some answers to (pb1) are available in $[3]$, $[4]$, $[5]$, $[9]$. For (pb2) we are able to give the following:

Theorem 2 ($[7]$, $[8]$) Let $f \in L^\infty(0, T)$; then the two-dimensional Lebesgue measure of the set of a. i. c. for which the uniqueness fails is zero.

4. The main tool for Theorem 2 is the following:

Theorem 3 ($[7]$, $[8]$) Suppose that $f \in L^\infty(0, T)$, that $\psi_h \in \text{Lip}(R) \cap L^\infty(R)$ for every h and that (2), (3) of Theorem 1 hold. Let $u^0 \in L^1(R^2) \cap L^\infty(R^2)$, $u^0 = 0$ for $x > 0$. Then the sequence $u_h(t, x, p)$, in $L^1((0, T) \times R^2) \cap L^\infty((0, T) \times R^2)$, of the solutions to the problems (I_h) $\dfrac{\partial u}{\partial t} + p\dfrac{\partial u}{\partial x} + (f(t) - \psi_h(x))\dfrac{\partial u}{\partial p} = 0$ on $[0, T] \times R^2$; $u(0, x, p) = u^0(x, p)$, has the following properties:
(j) $u_h \to u_\infty$ in $L^1((0, T) \times R^2)$;
(jj) for every $t \in [0, T]$, $u_h(t, ., .) \to u_\infty(t, ., .)$ in $L^1(R^2)$;
where $u_\infty \in L^1((0, T) \times R^2) \cap L^\infty((0, T) \times R^2)$ is zero for $x > 0$, but for $x \leq 0$ is the unique solution to the problem

$$(I_\infty) \quad \frac{\partial u}{\partial t} + p\frac{\partial u}{\partial x} + f(t)\frac{\partial u}{\partial p} = 0 \quad \text{on} \quad [0, T]xR^-xR;$$

$$u(0, x, p) = u^0(x, p) \, ; \quad u(t, 0, p) = u(t, 0, -p).$$

5.   Now we are able to formulate Theorem 1 giving prominence to the strict connection between the type of penalty method and the elastic nature of the limit problem.

<u>Theorem 1'</u>   Let $f \in L^\infty(0, T)$ and let (ψ_h) be a sequence as in Theorem 3. For almost all a. i. c. (x^0, p^0), if (x_h) is a sequence of solutions for (P_h), then such a sequence converges uniformly to a solution of the Cauchy's problem for the elastic bounce problem with (x^0, p^0) as a. i. c.

6.   OPEN PROBLEMS

(p1) Give a "reasonable" measure on $L^1(0, T)$ for which the measure of the set of functions $f \in L^1(0, T)$ for which the uniqueness fails is zero.

(p2) Give a penalization for the anelastic bounce problem, for a given $f(t)$.

(p3) Given $f \in L^\infty(0, T)$, study the structure of the set X of a. i. c. for which the uniqueness fails.   For example, it is interesting to prove or disprove the following conjecture: "there exists a rectangle R such that $\overline{R \cap X} = R$".

(p4) Prove Theorem 2 for $f \in L^1(0, T)$.

REFERENCES

1.   E. De Giorgi.   <u>Recent Methods in Nonlinear Analysis</u>, Roma, May 8/12-1978 (Eds. E. De Giorgi, E. Magenes, U. Mosco).

2.   E. De Giorgi.   Quelques problèmes de Γ-convergence, Proc. Meeting on <u>Computing Methods in Applied Sciences and Engineering</u>, Versailles, Dec. 1979, Ed. R. Glowinski, J. L. Lions, North-Holland (1980).

3.   M. Carriero, E. Pascali.   <u>Rend. Mat.</u> (4) <u>13</u>, serie 6 (1980).

4.   M. Carriero, E. Pascali.   <u>Boll. Un. Mat. Ital.</u> (6) <u>1-A</u> (1982).

5.   M. Carriero.   <u>Studio di Problemi-Limite dell'Analisi Funzionale</u>, Bressanone (1981).

6.   A. Leaci. Atti Acc. Lincei; Rend. Cl. Sc. Mat. Fis. Natur. 71 (1981).

7.   M. Carriero, A. Leaci, E. Pascali. Atti Acc. Lincei; Rend. Cl. Sc. Mat. Fis. Natur. 72 (1982).

8.   M. Carriero, A. Leaci, E. Pascali. Ann. Mat. Pura e Appl. (IV), CXXXIII (1983).

9.   D. Percivale. Preprint S. N. S. Pisa.

Michele Carriero and Eduardo Pascali
Dipartimento di Matematica
Università
Via Arnesano
I 73100 Lecce
Italy

E CAVAZZUTI
$\mathfrak{N}$-convergence and convergence of Nash equilibria

In this note we introduce a convergence for sequences of games which ensures that every limit of a convergent sequence of Nash equilibria is a Nash equilibrium for the limit game.

We call a (two-person) game a vector valued function $J : U_1 \times U_2 = U \to \widetilde{R}^2 = [-\infty, +\infty]^2$, where U_1 and U_2 are the strategic spaces for players 1 and 2, respectively. Here we consider only games on the product of strategic spaces, for simplicity.

Let us introduce the concept of Nash equilibrium (see, for example, [1]) by:

__Definition 1__   We say that $u^0 = (u_1^0, u_2^0) \in U$ is a Nash equilibrium point for the game $J = (J_1, J_2) \to \widetilde{R}^2$ if:

$$J_1(u_1^0, u_2^0) \leq J_1(u_1, u_2^0) \qquad \forall u_1 \in U_1$$

$$J_2(u_1^0, u_2^0) \leq J_2(u_1^0, u_2) \qquad \forall u_2 \in U_2$$

Let U_1, U_2 be two convergence spaces; when U_1, U_2 are topological we shall consider the convergence defined by the topology.

We denote by $\to$ the convergence on each space and on the product space; the case which occurs will be clear from the context. We define now a convergence for games, that will be called Nash convergence and denoted by $\underset{\to}{\mathfrak{N}}$.

__Definition 2__   We say that a sequence $\{J^h\}_h$ of games, $J^h : U \to \widetilde{R}^2$, $\mathfrak{N}$- converges to a game J^0 if:

$$\forall u_1^0, \forall u_2^0; \ \forall u_1^h \to u_1^0, \ \forall u_2^h \to u_2^0$$

$$\underline{\lim}_h J_i^h(u_1^h, u_2^h) \geq J_i^0(u_1^0, u_2^0) \qquad i = 1, 2 \qquad (1)$$

92

$$\forall u_1^0, \; \forall u_2^0 \, ; \; \forall u_2^h \to u_2^0 \; \exists \, u_1^h \to u_1^0 \; :$$

$$\overline{\lim}_h J_1^h(u_1^h, \, u_2^h) \leq J_1^0(u_1^0, \, u_2^0) \tag{2}$$

$$\forall u_1^0, \; \forall u_2^0 \, ; \; \forall u_1^h \to u_1^0 \; \exists \, u_2^h \to u_2^0 \; :$$

$$\overline{\lim}_h J_2^h(u_1^h, \, u_2^h) \leq J_1^0(u_1^0, \, u_2^0) \tag{3}$$

We can prove the following results:

__Theorem 1__ (Convergence)   Let a sequence of games $\{J^h\}_h$ be given and the following conditions satisfied:

(i) $J^h \underset{\to}{\mathfrak{N}} J^0$

(ii) u^h is a Nash equilibrium for J^h

(iii) $u^h \to u^0$.

Then we have:

(a) u^0 is a Nash equilibrium for J^0 ;

(b) $J^0(u^0) = \lim_h J^h(u^h)$.

__Theorem 2__ (Stability for perturbation)   Let $\{J^h\}_h$ be given and $I_0 : U \to R^2$ be a continuous function. If $J^h \underset{\to}{\mathfrak{N}} J^0$, then

$$(J_h + I_0) \underset{\to}{\mathfrak{N}} (J_0 + I_0)$$

Every result stated remains valid for n-person games.

Characterizations of $\mathfrak{N}$-convergence by means of the Γ-limits of De Giorgi are possible in first countable topological spaces. For more details we refer to [2] and [3].

REFERENCES

1.   J. P. Aubin. Mathematical Methods of Game and Economic Theory, North Holland (1979).

2.   E. Cavazzuti, N. Pacchiarotti. Convergence of games and Nash equilibria (to appear).

3.   E. De Giorgi. Γ-convergenza e G-convergenza, Boll. Un. Mat. Ital. (5) 14-A (1977) 213-220.

Ennio Cavazzuti
Istituto Matematico
Università
Via Campi, 213/B
I 41100 Modena
Italy

R DE ARCANGELIS
A problem of Γ-convergence in weighted Sobolev spaces

In this communication I want to describe a result dealing with the approximability of degenerate quadratic functionals, defined on weighted Sobolev spaces, by functionals connected with operators of isotropic type.

If $\Lambda \geq 1$ let $M(\Lambda)$ be the class of all symmetric function matrices (a_{ij}) measurable on R^n verifying

$$|z|^2 \leq \sum_{ij} a_{ij}(x) z_i z_j \leq \Lambda |z|^2$$

for every $i, j \in \{1, \ldots, n\}$, x a.e. in R^n, $z \in R^n$.

Moreover let $W(R^n)$ be the set of those functions w such that

$$w(x) > 0 \quad \text{a.e. in} \quad R^n, \quad w, w^{-1} \in L^1_{loc}(R^n) \ .$$

If Ω is a bounded open set in R^n and $w \in W(R^n)$, $H^1_0(\Omega, w)$ will be the weighted Sobolev space defined as the completion of $C^1_0(\Omega)$ in the topology induced by the norm $\|u\|_{H^1_0(\Omega, w)} = (\int_\Omega |Du|^2 w)^{\frac{1}{2}}$ and $H^{-1}(\Omega, w)$ will be its dual.

We explicitly observe that the summability of w^{-1} ensures the continuous embedding of $H^1_0(\Omega, w)$ in $H^{1,1}_0(\Omega)$ and the compact one in $L^1(\Omega)$.

Using techniques of Γ-convergence (see [3]), in [2] is proved the following result that extends to the case of degenerate operators a result of A. Marino and S. Spagnolo (see [4]):

<u>Theorem</u>   Let $(a_{ij}) \in M(\Lambda)$, then there exists a sequence of functions $(\beta_h)_h$, with $c^{-1} \leq \beta_h(x) \leq c\Lambda$ for every x (c depending only on the dimension n), such that for every $w \in W(R^n)$, for every Ω bounded open set in R^n and for every $\psi \in H^{-1}(\Omega, w)$ the sequence $(u_h)_h$ of the solutions of the problems

$$\underset{u \in H_0^1(\Omega, w)}{\text{Min}} \{ \int_\Omega \beta_h(x) \, w(x) \, |Du|^2 \, dx + \langle \psi, u \rangle \}$$

converges in $L^1(\Omega)$ to the solution u_0 of the problem

$$\underset{u \in H_0^1(\Omega, w)}{\text{Min}} \{ \int_\Omega \sum_{ij} a_{ij}(x) \, w(x) \, D_i u D_j u \, dx + \langle \psi, u \rangle \}.$$

Let us explicitly observe that the functions β_h are independent of w.

This theorem might be considered as a stability result in $\Gamma^-(L^1)$ - convergence with regard to the multiplication of the integrands of functionals of the type $\int_\Omega \sum_{ij} a_{ij}(x) \, D_i u D_j u$, where (a_{ij}) is in $M(\Lambda)$, for a function w in $W(R^n)$.

Stability problems of this kind are treated in [1] in the case w continuous and in [2] in the case $w \in W(R^n)$ with $n = 1$. Nevertheless there exist counter-examples (see [1]) even in the case when w is bounded, but with not summable inverse.

To show the stated theorem, it is necessary first to examine the $\Gamma^-(L^1)$ - convergence of functionals of the kind considered in the one-dimensional case, then that of those functionals whose matrix of the coefficients is diagonal with the element in the (i, i)-position written as the product of two functions, the first depending only on the i-th variable and the latter depending on the remaining ones.

For these kinds of functionals, stability results, in the sense stated before, are given.

REFERENCES

1. L. Carbone, C. Sbordone. Some properties of Γ-limits of integral functionals, Ann. Mat. Pura Appl. (4) 122 (1979) 1-60.

2. R. De Arcangelis. Sulla G-approssimabilita di operatori ellittici degeneri in spazi di Sobolev con peso, Rend. Mat. (to appear).

3. E. De Giorgi, T. Franzoni. Su un tipo di convergenza variazionale, Atti Accad. Naz. Lincei, Rend. Cl. Sci. Mat. Fis. Natur. (8) 58 (1975) 842-850.

4.   A. Marino, S. Spagnolo. Un tipo di approssimazione dell'operatore $\sum_{ij} D_i(a_{ij} D_j)$ con operatori $\sum_j D_j(\beta D_j)$, <u>Ann. Sc. Norm. Sup. Pisa,</u> <u>Cl. Sci.</u> (3) <u>23</u> (1969) 657-673.

Riccardo De Arcangelis
Istituto di Matematica
Università
Via Mezzocannone, 8
I 80134 Napoli
Italy

M DEGIOVANNI
Some operators with lack of monotonicity

In a joint work with E. De Giorgi, A. Marino and M. Tosques [2] we have introduced a class of nonlinear operators which extends that of Lipschitz continuous perturbations of maximal monotone operators [1] and seems useful in dealing with evolution equations with non-convex unilateral constraints.

Of the results proved in [3], the main result is given here.

Let Ω be an open subset of a real Hilbert space H whose scalar product and norm are denoted by $(\cdot, \cdot)$ and $|\cdot|$.

Let $f : \Omega \to \mathbb{R} \cup \{+\infty\}$ be a lower semicontinuous function and $A : H \to \mathcal{P}(H)$ an operator such that

$$D(A) = \{u \in H : Au \neq \emptyset\} \subset D(f) = \{u \in \Omega : f(u) \in \mathbb{R}\}.$$

We set

$$|A_0 u| = \begin{cases} \inf\{|\alpha| : \alpha \in Au\} & \text{if } u \in D(A) \\ +\infty & \text{if } u \in H \backslash D(A) . \end{cases}$$

<u>Definition</u> If $\phi : \Omega \times \mathbb{R}^2 \to \mathbb{R}^+$ is a continuous function, the operator A is said to be (ϕ, f)-monotone if

$$(\alpha - \beta | u - v) \geq -[\phi(u, f(u), |\alpha|) + \phi(v, f(v), |\beta|)] |u - v|^2$$

whenever $u, v \in D(A)$, $\alpha \in Au$, $\beta \in Av$.

<u>Definition</u> The operator A is said to be f-solvable at a point u of $D(A)$ if there exist

$$c_0 > f(u), \quad c_1 > |A_0 u|, \quad M \geq 0, \quad r > 0, \quad \lambda_0 > 0$$

such that

(a) $\forall \lambda$ in $]0, \lambda_0]$, $\forall v$ in $B(u, r) \cap D(A)$ with $f(v) \leq c_0$, $|A_0 v| \leq c_1$, $\exists w$ in $D(A)$:

$$\frac{v - w}{\lambda} \in Aw, \quad \frac{|v - w|}{\lambda} \leq M, \quad f(w) \leq f(v) + \lambda M ;$$

(b) for every sequence $(v_n)_n$ in $D(A)$ converging to v in $B(u, r)$ with $f(v_n) \leq c_0$, for every sequence $(\alpha_n)_n$ weakly convergent to α with $\alpha_n \in Av_n$, $|\alpha_n| \leq c_1$, we have $v \in D(A)$ and $\alpha \in Av$.

<u>Proposition</u>   Let A be a (ϕ, f) -monotone operator which is f-solvable at a point u of $D(A)$. Then Au has a unique element of minimal norm which we denote by $A_0 u$.

<u>Theorem</u>   Let A be a (ϕ, f) -monotone operator which is f-solvable at every point of $D(A)$. Then, for every u in $D(A)$, there exists $T > 0$ and a unique curve $U : [0, T[\rightarrow D(A)$ such that $U(0) = u$ and such that

(a) U is Lipschitz continuous on $[0, T[$;

(b) $f \circ U$ is bounded on $[0, T[$;

(c) for every t in $[0, T[$ there exists the right derivative $U'_+(t)$ and

$$U'_+(t) = -A_0 U(t) .$$

<u>REFERENCES</u>

1.   H. Brézis, <u>Opérateurs Maximaux Monotones,</u> Notes de Mathematica (50) , North-Holland (1973) .

2.   E. De Giorgi, M. Degiovanni, A. Marino, M. Tosques, Evolution equations for a class of nonlinear operators, <u>Atti Accad. Naz. Lincei Rend. Cl. Sci. Fis. Mat. Natur.</u> (8) <u>75</u> (1983) 1-8.

3.   M. Degiovanni, M. Tosques, Evolution equations for (ϕ, f) -monotone operators, <u>Boll. Un. Mat. Ital. B</u> (in press) .

Marco Degiovanni
Scuola Normale Superiore, Piazza dei Cavalieri, 7
I 56100 Pisa, Italy

F DONATI

The solutions of some semilinear Dirichlet problems

In this note we consider problems of the type

$$(P) \quad \begin{cases} \Delta u + g(u) = h & \text{in } \Omega \\ u = 0 & \text{on } \partial\Omega \end{cases}$$

where Ω is a bounded, smooth domain of $\mathbb{R}^n$, $h \in C^{0,\alpha}(\bar{\Omega})$ $(0 < \alpha < 1)$ and $g \in C^2(\mathbb{R})$, $g(0) = 0$ with finite $\lim_{r \to \pm\infty} g'(r) = g'(\pm\infty)$.

When g also satisfies $0 \le g'(0) < \lambda_1 < g'(\pm\infty) < \lambda_2$ and $g''(r)r > 0$ for $r \ne 0$ (λ_i, $i = 1, 2$, being the i-th eigenvalue of Δ) then it is known that (see e.g. [3], [5]) for every $h \in C^{0,\alpha}(\bar{\Omega})$ problem (P) has a solution $u \in C_0^{2,\alpha}(\bar{\Omega})$ and there exists $R > 0$ such that if $\|h\|_{0,\alpha} < R$ then (P) has exactly three solutions. Under the same assumptions for g we proved in [2] that (P) has a unique solution at infinite in the sense of the following:

<u>Theorem 1</u> For all $q \in C^{0,\alpha}(\bar{\Omega})$ such that $(q, \phi_1)_{L^2} = 0$, where ϕ_1 is the first (positive) eigenfunction of Δ, there exist $t_+ = t_+(q) > 0$ and $t_- = t_-(q) < 0$ such that, if $h = t\phi_1 + q$ with $t > t_+$ [$t < t_-$], then problem (P) has a unique solution which is positive [negative].

The techniques used to prove the above statement also allowed us to obtain further information about the solutions of (P), that is:

<u>Theorem 2</u> Let $h = t\phi_1$ in (P), then:

(i)   there exist $\bar{t}, \underline{t} \in \mathbb{R}$, $\bar{t} > 0 > \underline{t}$, such that for $0 < t < \bar{t}$ [$\underline{t} < t < 0$] there are exactly two negative [positive] solutions; for $t = \bar{t}$ [$t = \underline{t}$] there is a unique negative [positive] solution and for $t = 0$ there exist exactly three solutions u_-, 0, u_+ such that $u_- < 0 < u_+$ in Ω;

(ii)   there exists $\varepsilon \in \mathbb{R}$, $0 < \varepsilon < \min\{\bar{t}, -\underline{t}\}$, such that, for

100

$0 < t < \varepsilon$ [$-\varepsilon < t < 0$], there exist exactly three solutions, one positive [negative] and two negative [positive].

The main rool for the proofs of Theorems 1 and 2 was the splitting of the original problem into two problems satisfying a suitable variant of the well-known result of Ambrosetti and Prodi [1].

We remark that the results here stated still hold true when the nonlinearity g satisfies more general assumptions and that a geometric description of the singular manifold of the map $\Phi = \Delta + g$ can also be given, see [4].

REFERENCES

1. A. Ambrosetti and G. Prodi, On the inversion of some differentiable mappings with singularities between Banach spaces, Ann. Mat. Pura Appl. 93 (1972) 231-246.

2. V. Cafagna and F. Donati, On the sign of the solutions to some semilinear Dirichlet problems. Preprint.

3. A. Castro and A. C. Lazer, Critical point theory and the number of solutions of a nonlinear Dirichlet problem, Ann. Mat. Pura Appl. 120 (1979) 113-137.

4. F. Donati, Some properties of the solutions to a class of semilinear Dirichlet problems (in preparation).

5. J. K. Kazdan and F. W. Warner, Remarks on some quasilinear elliptic equations, Comm. Pure Appl. Math. 28 (1975) 567-597.

Flavio Donati
Istituto di Matematica
Università
Via Roma, 33
I 67100 L'Aquila
Italy

P DONATO
Weak convergence of non-uniformly oscillating functions

In analysis we often need to consider the behaviour, for $\varepsilon \to 0$, of function sequences of the form $a^\varepsilon(x) = a(x/\psi(\varepsilon), x/\varepsilon)$, $x \in R^N$, with $\psi(\varepsilon)$ convergent for $\varepsilon \to 0$ and $a(x, y)$ periodic in y if $\lim_{\varepsilon \to 0} \psi(\varepsilon) \neq 0$, in (x, y) otherwise. In this communication we examine the minimal hypothesis of regularity in order to obtain the weak convergence of $a^\varepsilon(x)$ to a properly defined "mean value".

For brevity, only the case $\psi \equiv 1$, i.e. $a^\varepsilon(x) = a(x, x/\varepsilon)$, is considered. For the other cases, for references and details we refer to $[1]$ where some examples showing the limits of possible extensions are also given.

Let $p \in [1, +\infty]$. If $a \in L^p_{loc}(R^N)$ we say that a has a __mean value__ $M(a)$ if, for all bounded Borel sets B, there exists

$$\lim_{T \to +\infty} \frac{1}{|TB|} \int_{TB} a(x)\, dx$$

independent of B. We define $M^p = \{a \in L^p_{loc}(R^N) : \exists M(a)$ and $\forall \omega \subset\subset R^N \frac{1}{T} \int_{T\omega} |a(y)|^p dy \leq C$, for T large, if $p < +\infty\}$. It is intrinsic in the definition that if $a \in M^p$ then $a^\varepsilon \to M(a)$, for $\varepsilon \to 0$, weakly in $L^p_{loc}(R^N)$ if $p < +\infty$, in $L^\infty_{loc}(R^N)$ weakly $\star$ if $p = +\infty$.

Let us note that periodic functions and, more generally, almost periodic functions in the sense of Besicovitch, belong to M^p.

Let Ω be an open set of R^N and let us define the following classes:

$$\Sigma = \{f : \Omega \times R^N \to R : \forall \omega \subset\subset \Omega,\ f(x, y) = \sum_{\text{finite}} \chi_i(x) f_i(y),$$

$$f_i \in M^\infty,\ \cup S_i \supset \omega,\ S_i \text{ half-open intervals},\ S_i \cap S_j = \emptyset,$$

$$\chi_i \text{ characteristic function of } S_i\},$$

and, for $1 \leq p \leq +\infty$:

$$K^p = \{ a \in \Omega \times R^N \to R : \forall \omega \subset\subset \Omega \| a^\varepsilon \|_{L^p(\omega)} \le c \text{ and}$$

there exists $\{ a_h \} \subset \sum$ s.t. $\forall \omega \subset\subset \Omega \| a_h \|_{L^1(\omega)} \le c$

and

(H_1) $\displaystyle \lim_h \lim_\varepsilon \sup \left| \int_B [a(x, x/\varepsilon) - a_h(x, x/\varepsilon)] dx \right| = 0$

for all bounded Borel sets $B \subset \Omega \}$.

<u>Proposition</u> If $a \in K^p$ then there exists $\tilde{a} \in L^p_{loc}(\Omega)$ such that $a^\varepsilon \rightharpoonup \tilde{a}$, weakly in $L^p_{loc}(\Omega)$ if $p < +\infty$, in $L^\infty_{loc}(\Omega)$ weakly $\star$ if $p = +\infty$. Moreover, if the following condition holds:

(H_2) $\begin{cases} a(x, \cdot) \in M^1 \ \forall x \in \Omega \text{ and there exists } \{ a_h \} \subset \sum \text{ satisfying} \\ (H_1) \text{ and s.t. } M_y(a_h) \to M_y(a) \text{ weakly in } L^1_{loc}(\Omega) \end{cases}$

then $\tilde{a}(x) = M_y(a)$.

<u>Corollary</u> If $a(x, \cdot)$ is Y-periodic $\forall x \in \Omega$, then $a^\varepsilon \to M_y(a)$ if $a(x, y) \in C^0(R^N_y, L^\infty_{loc}(\Omega))$ or if $a(\cdot, y) \in C^0(\Omega)$ for a.e. y and $\| a(\cdot, y) \|_{C^0(\bar\omega)} \in L^p(Y)$, $\forall \omega \subset\subset \Omega$ (in particular if $a \in C^0(\Omega, L^\infty(R^N_y))$).

The following example is a variant of one contained in [1]. It shows a function $a(s, y)$ defined almost everywhere on R^2, not verifying (H_2), for which the weak limit of a^ε is different from $M(a)$.

<u>Example</u> We set, for $k \in N$, $B_k = \{ (x, y) \in ([0, 1])^2 : x = ky \pm c, |c| < \eta_k \}$ and $B = \bigcup_{k \in N} B_k$. We have that $meas(B_k) < 2\sqrt{2}\eta_k$, and, choosing $\eta_k = 1/4\sqrt{2} \, 2^k$, we have $meas(B) < \frac{1}{2}$. If we set for $(x, y) \in ([0, 1])^2$

$a(x, y) = \begin{cases} 1 \text{ if } (x, y) \in B \\ 0 \text{ otherwise} \end{cases}$

and we prolong a for periodicity, we have that for any $\hat{a}$ in the same

103

equivalence class of a it is possible to find a sequence $\hat{\varepsilon}_k \to 0$ such that $\hat{a}(x, x/\hat{\varepsilon}_k) \equiv 1$, but $M(\hat{a}) \leq \frac{1}{2}$ since $\text{meas}\{(x, y) \in ([0, 1])^2 / a(x, y) = 0\} \geq \frac{1}{2}$.

REFERENCES

1.   P. Donato. Alcune ossrvazioni sulla convergenza debole di funzioni non uniformemente oscillanti, Ric. di Mat. 2 (1984).

Patrizia Donato
Istituto di Matematica
Università
Via Mezzocannone, 8
I 80134 Napoli
Italy

M G GARRONI
Green's function and asymptotic behaviour of the solution of some oblique derivative problem not in divergence form

1.   Let Ω be a bounded open subset of $\mathbb{R}^N$, $N \geq 2$, with boundary Γ of class C^2. We shall denote by Q_T the cylinder $\Omega \times \,]0,\,T[$, $0 < T < +\infty$, and by $\Sigma_T = \Gamma \times \,]0,\,T[$ its lateral boundary.

We set

$$(1) \qquad A = - \sum_{i,j=1}^{N} a_{ij}(x)\, \frac{\partial^2}{\partial x_i \partial x_j} + \sum_{i=1}^{N} a_i(x)\, \frac{\partial}{\partial x_i}\,, \quad B = \sum_{i=1}^{N} b_i(x)\, \frac{\partial}{\partial x_i}$$

whose coefficients are assumed to satisfy

$$(2) \quad \left|
\begin{array}{ll}
\text{(i)} & a_{ij},\ a_i,\ b_i \in C^{0,\alpha}(\Omega),\ \text{for some}\ 0 < \alpha < 1 \\[2mm]
\text{(ii)} & \displaystyle\sum_{i,j=1}^{N} a_{ij}(x)\,\xi_i\xi_j \geq \mu\,|\xi|^2,\ \mu > 0,\ \forall \xi \in \mathbb{R}^N,\ \forall x \in \Omega \\[2mm]
\text{(iii)} & \displaystyle\left|\sum_{i=1}^{N} b_i(x)\,\nu_i(x)\right| \geq \beta > 0,\ \forall x \in \Gamma.
\end{array}
\right.$$

Here $\nu = (\nu_1,\, \nu_2,\, \ldots,\, \nu_N)$ is the outward normal vector to Γ.

<u>Theorem 1</u>   Under the assumptions made there exists the Green's function $G \equiv G(x,\, y,\, t,\, \tau)$ of the initial boundary value problem:

$$(3) \quad \left\{
\begin{array}{l}
\dfrac{\partial G}{\partial t} + A_x G = \partial(x - y)\,\partial(t - \tau),\ \text{in}\ Q_T \\[3mm]
G(x,\, y,\, t,\, \tau)\big|_{t \leq \tau} = 0 \\[3mm]
B_x G = 0,\ \text{on}\ \Sigma_T\,.
\end{array}
\right.$$

G satisfies the estimates

$$(4) \begin{cases} |G(x,\ y,\ t,\ \tau)| \le c_1(t-\tau)^{-\frac{N}{2}} e^{-c_2 \frac{|x-y|^2}{t-\tau}} \\[2mm] |\frac{\partial}{\partial x_i} G(x,\ y,\ t,\ \tau)| \le c_1(t-\tau)^{-\frac{N+1}{2}} e^{-c_2 \frac{|x-y|^2}{t-\tau}} \\[2mm] |\frac{\partial}{\partial t^s} \frac{\partial}{\partial t^\tau} G(x,\ y,\ t,\ \tau)| \le c_1(t-\tau)^{-\frac{N+1+\alpha}{2}} (\rho^{\alpha-1}(x) V(t-\tau)^{\frac{\alpha-1}{2}}) e^{-c_2 \frac{|x-y|^2}{t-\tau}} \end{cases}$$

where $2s + \tau = 2$; s, $\tau \in \mathbb{N}$ and $\rho(x) = \text{dist}(x,\ \Gamma)$. Moreover, we have

$$(5) \begin{cases} G(x,\ y,\ t,\ \tau) \ge 0 \\[1mm] \text{there exists a ball } B \subset \Omega \text{ and a constant } \delta \text{ such that} \\[1mm] G(x,\ y,\ t,\ \tau) \ge \delta > 0,\ \forall x \in \Omega,\ \forall y \in B,\ \forall t > \tau \\[1mm] \int_\Omega G(x,\ y,\ t,\ \tau)\, dy = 1 \end{cases}$$

<u>Theorem 2</u>   For every $\lambda > 0$, and $f \in L^p(\Omega)$, $(1 < p < \frac{1}{1-\alpha})$, the oblique derivative problems

$$(6) \begin{cases} u_\lambda \in W_p^2(\Omega) \\[1mm] Au_\lambda + \lambda u_\lambda = f \text{ in } \Omega,\ Bu_\lambda = 0 \text{ on } \Gamma; \end{cases}$$

and

$$(7) \begin{cases} v_\lambda \in W_p^2(\Omega),\ \text{Max}[v_\lambda,\ Av_\lambda + \lambda v_\lambda - f] = 0 \text{ in } \Omega \\[1mm] Bv_\lambda = 0 \text{ on } \Gamma \end{cases}$$

are unique solutions of u_λ and v_λ respectively; we have also the **representation**

$$(8) \qquad u_\lambda(x) \equiv (\Gamma_\lambda f)(x) = \int_0^{+\infty} e^{-\lambda t} [\int_\Omega G(x,\ y,\ t,\ 0) f(y)\, dy]\, dt.$$

2.   <u>FREDHOLM ALTERNATIVE AND INVARIANT MEASURE</u>

We are interested in the asymptotic behaviour of u_λ and v_λ when $\lambda \to 0$. We consider the following problems:

106

(9) $u \in L^p(\Omega)$, $(I - \lambda J_\lambda) u = J_\lambda f$

(10) $m \in L^{p'}(\Omega)$, $(I - \lambda J_\lambda^\star) m = 0$ $\dfrac{1}{p} + \dfrac{1}{p'} = 1$

where J_λ is defined by (8) and $J_\lambda^\star$ is its adjoint.

<u>Theorem 3</u>   For every $\lambda > 0$, equation (10) has a unique solution m such that

$$m > 0, \quad \frac{1}{|\Omega|} \int m dx = 1 .$$

Moreover,

$$\int_\Omega \left[\int_\Omega G(x, y, t, 0) f(y) dy \right] m(x) dx = \int_\Omega f(x) m(x) dx, \quad \forall t > 0, \ \forall f \in L_p(\Omega),$$

$$\left\| \int_\Omega G(x, t, 0) f(y) dy - \frac{1}{|\Omega|} \int fm \, dy \right\|_{L^p} \leq K \|f\|_p e^{-\rho t}, \quad \forall t > 0, \ \forall f \in L^p(\Omega),$$

with k, ρ independent of f.

We can define probability transition functions P by setting

$$P(x, t, E) = \int_E G(x, y, t, 0) dy, \quad \text{Borel subset } E \text{ of } \overline{\Omega} .$$

These are related to a diffusion process reflected at the boundary Γ according to the vector field B, and $d\mu = \dfrac{1}{|\Omega|} m dx$ is the invariant measure of this process.

3.   ASYMPTOTIC BEHAVIOUR

The behaviour of the solutions u_λ and v_λ and the form of the limit problems depend on $\int_\Omega fm dx$ where m is given by Theorem 3.

<u>Theorem 4</u>   Let $f \in L^p$ $(1 < p < \dfrac{1}{1-\alpha})$, then $\|u_\lambda\|_{W_p^2}$ is bounded by a constant independent of λ if and only if $\int_\Omega fm dx = 0$. In this case $\lim\limits_{\lambda \to 0} u_\lambda = u$ in W_p^2 weakly and

(11) $u(x) = \int_0^\infty (\int_\Omega G(x, y, t, 0) f(y) dy) dt$

107

is the unique solution such that $\int_\Omega u m dx = 0$.

(i)   If $\int_\Omega f m dx > 0$, v_λ converge weakly in $W_p^2(\Omega)$ to the unique solution v of

(12) $\{ v \in W_p^2(\Omega), \ \text{Max}[v; Av-f] = 0 \ \text{in} \ \Omega, \ Bv = 0 \ \text{on} \ \Gamma$

(ii)   If $\int_\Omega f m dx < 0$, then $w_\lambda = v_\lambda - \frac{1}{|\Omega|} \int_\Omega v_\lambda m dx$ converge weakly in $W_p^2(\Omega)$ to the unique solution w of

(13) $\begin{cases} w \in W_p^2(\Omega), \ Aw = f - \frac{1}{|\Omega|} \int_\Omega f m dx \ \text{in} \ \Omega, \ Bw = 0 \ \text{on} \ \Gamma, \\ \int_\Omega w m dx = 0. \end{cases}$

(iii)   If $\int f m dx = 0$, $p > \frac{N}{2}$, then v_λ converge weakly in $W_p^2(\Omega)$ to the unique solution v of

(14) $v \in W_p^2(\Omega), \ v \leq 0, \ Av = f \ \text{in} \ \Omega, \ Bv = 0 \ \text{on} \ \Gamma.$

The previous results are contained in a more general setting in [1] and [2].

These results can be extended to some integrodifferential operators, corresponding to diffusions with jumps (see [3]).

REFERENCES

1.   M. G. Garroni and V. A. Solonnikov. On parabolic oblique derivative problems with Hölder continuous coefficients, Comm. on P. D. E. (to appear).

2.   I. Capuzzo Dolcetta and M. G. Garroni. Fredholm alternative and invariant measures for some oblique derivative problems not in divergence form (to appear).

3.   M. G. Garroni and K. Menaldi. Invariant measures for integro-differential problems (to appear).

Further references on the subject can be found in the papers mentioned above.

Maria Giovanna Garroni
Dipartimento di Matematica
Università "La Sapienza"
Piazzale A. Moro, 5
I 00185 Roma,
Italy

108

M MARINO & A MAUGERI

L^p theory and partial Hölder continuity for quasilinear parabolic systems of higher order with strictly controlled growth

This note concerns some regularity results for solutions of the quasilinear parabolic system of order $2m$:

$$(-1)^m \sum_{|\alpha|=m} \sum_{|\beta|=m} D^\alpha(A_{\alpha\beta}(X, \delta u) D^\beta u) + \frac{\partial u}{\partial t}$$

$$= (-1)^m \sum_{|\alpha|=m} D^\alpha f^\alpha(X, \delta u) + \sum_{|\alpha|\leq m-1} (-1)^{|\alpha|} D^\alpha f^\alpha(X, Du)$$

where $u : Q \to \mathbb{R}^N$ $(Q = \Omega \times (-T, 0))$, $\delta u = \{D^\alpha u\}_{|\alpha|\leq m-1}$, $Du = \{D^\alpha u\}_{|\alpha|\leq m}$ and $X = (x, t)$.

We consider the solution $u \in L^2(-T, 0, H^m(\Omega, \mathbb{R}^N)) \cap L^\infty(-T, 0, L^2(\Omega, \mathbb{R}^N))$ such that

$$\int_Q \{ \sum_{|\alpha|=m} \sum_{|\beta|=m} (A_{\alpha\beta} D^\beta u | D^\alpha \psi) - (u | \frac{\partial \psi}{\partial t}) \} dX = \int_Q \sum_{|\alpha|\leq m} (f^\alpha | D^\alpha \psi) dX$$

$$\forall \psi \in L^2(-T, 0, H_0^m(\Omega, \mathbb{R}^N)) \cap H^1(-T, 0, L^2(\Omega, \mathbb{R}^N)) : \psi(x, T)$$

$$= \psi(x, 0) = 0 \text{ in } \Omega$$

and suppose that the vectors f^α of $\mathbb{R}^N$ have the following strictly controlled growth:

$$\|f^\alpha(X, \delta u)\| \leq g^\alpha(X) + c \sum_{|\beta|\leq m-1} \|D^\beta u\|^{\theta(m, |\beta|)}, \quad |\alpha| = m,$$

$$\|f^\alpha(X, Du)\| \leq g^\alpha(X) + c \sum_{|\beta|\leq m} \|D^\beta u\|^{\theta(|\alpha|, |\beta|)}, \quad |\alpha| \leq m-1,$$

with

$$1 \leq \theta(m, |\beta|) < \frac{n+2m}{n+2|\beta|}, \quad |\beta| \leq m-1,$$

$$1 \leq \theta(|\alpha|, |\beta|) < \frac{n+4m-2|\alpha|}{n+2|\beta|}, \quad |\alpha| \leq m-1, |\beta| \leq m.$$

109

We also suppose that $A_{\alpha\beta}(X, p^\star)$, $|\alpha| = |\beta| = m$, are $N \times N$ matrices, uniformly continuous and bounded in $\bar{Q} \times \mathcal{R}^\star$ $(\mathcal{R}^\star = \prod\limits_{|\alpha| \leq m-1} R_\alpha^N)$ and such that

$$\sum_{|\alpha|=m} \sum_{|\beta|=m} (A_{\alpha\beta}(X, p^\star) \xi^\beta | \xi^\alpha) \geq \nu \sum_{|\alpha|=m} \| \xi^\alpha \|^2, \quad \nu > 0,$$

for every $(X, p^\star) \in \bar{Q} \times \mathcal{R}^\star$ and for every system $\{\xi^\alpha\}_{|\alpha|=m}$ of vectors of $\mathbb{R}^N$.

The case $m = 1$ has been studied by S. Campanato under the assumption that f^α have strictly controlled growth and by Giaquinta-Struwe under the assumption that f^α have quadratic growth (see [2] for the references); if $m > 1$ and f^α have linear growth, partial Hölder continuity results for the derivatives $D^\alpha u$, $|\alpha| = m-1$, are shown in [1].

In the present work, using the technique of regularization in the space $\mathcal{L}^{(p,\theta)}(Q, \delta)$, we show the following

Theorem 1.1   If

$$g^\alpha(X) \in L^p(Q), \quad |\alpha| = m,$$

$$g^\alpha(X) \in L^{\frac{p}{\bar{\gamma}|\alpha|}}(Q), \quad |\alpha| \leq m-1,$$

with $p > n + 2m$, $\bar{\gamma}_{|\alpha|} = \dfrac{n+4m-2|\alpha|}{n + 2m}$, there exists a set $Q_0 \subset Q$, closed in Q, such that

$$D^\alpha u \in C^{0,\gamma}(Q \setminus Q_0, \mathbb{R}^N), \quad \forall \gamma < 1 - \frac{n + 2m}{p}, \quad |\alpha| = m-1$$

and

$$\mathcal{M}_{n+2m-2}(Q_0) = 0$$

where $\mathcal{M}_{n+2m-2}$ is the $(n + 2m - 2)$-dimensional Hausdorff measure with respect to the parabolic metric

$$\delta(X, Y) = \max \{\|x-y\|, |t-\tau|^{\frac{1}{2m}}\}, \quad X = (x, t), \quad Y = (y, \tau).$$

We obtain this result by using the following L^p regularity result:

__Theorem 1.2__   Suppose that $A_{\alpha,\beta}(X, \delta u) = A_{\alpha\beta}(X)$, $|\alpha| = |\beta| = m$. There exist $\tilde{\gamma}$, γ_h, $h = 0, 1, \ldots, m-1$, and q with

$$\max(\frac{n-2}{2}, \frac{m-1}{m}) < \tilde{\gamma} < 1 ,$$

$$\frac{2}{\tilde{\gamma}_h} < \frac{2}{\gamma_h} \leq \min\{\frac{2(n+2m)}{(n+2|\beta|)\theta(h, |\beta|)} , |\beta| \leq m\}, h = 0, 1, \ldots, m-1,$$

$$2 < q \leq \frac{2}{\tilde{\gamma}} \wedge \min\{\frac{2(n+2m)}{(n+2|\beta|)\theta(m, |\beta|)}, |\beta| \leq m-1\}$$

such that, if

$$g^{\alpha}(X) \in L^q(Q) , \quad |\alpha| = m,$$

$$g^{\alpha}(X) \in L^{\frac{2}{\gamma|\alpha|}}(Q), \quad |\alpha| \leq m-1 ,$$

then $D^{\alpha}u \in L^q_{loc}(Q, \mathbb{R}^N)$, $|\alpha| = m$, and, for any $Q(X^0, 2\sigma) \subset\subset Q$, we have

$$(\int_{Q(X^0, \sigma)} \sum_{|\alpha|=m} \|D^{\alpha}u\|^q dX)^{\frac{1}{q}} \leq c(\int_{Q(X^0, 2\sigma)} \sum_{|\alpha|=m} |g^{\alpha}|^q dX)^{\frac{1}{q}}$$

$$+ c \sum_{h=0}^{m-1} \sigma^{m-h+(n+2m)(\frac{1}{q}-\frac{\gamma_h}{2})} (\int_{Q(X^0, 2\sigma)} \sum_{|\alpha|=h} |g^{\alpha}|^{\frac{2}{\gamma_h}} dX)^{\frac{\gamma_h}{2}}$$

$$+ c(u) \sigma^{(n+2m)(\frac{1}{q}-\frac{1}{2})} [\int_{Q(X^0, 2\sigma)} (1 + \sum_{|\beta| \leq m} \|D^{\beta}u\|^{\frac{2(n+2m)}{n+2|\beta|}}$$

$$+ \sigma^{-2m} \|u - P(u, X^0, 2\sigma; x)\|^2) dX]^{\frac{1}{2}} ,$$

where $P(u, X^0, \sigma; x)$ is the vector polynomial in x, of degree at most $m - 1$, such that:

$$\int_{Q(X^0, \sigma)} D^{\alpha}(u - P) dX = 0 \quad \forall \alpha, |\alpha| \leq m-1 .$$

Finally we recall that, when we consider parabolic systems of order $2m$, the Poincaré's inequality that we need is the following:

<u>Theorem 1.3</u>   If $u \in L^2(-\sigma^{2m}, 0, H^m(B(\sigma), \mathbb{R}^N)) \cap H^{\frac{1}{2}}(-\sigma^{2m}, 0, L^2(B(\sigma), \mathbb{R}^N))$, then, for every integer j with $0 \le j \le m-1$, we have

$$\int_{Q(\sigma)} \sum_{|\alpha|=j} \| D^\alpha(u-P_{0,\sigma}) \|^2 dX \le c\sigma^{2m-2j} \{ \int_{Q(\sigma)} \sum_{|\alpha|=m} \| D^\alpha u \|^2 dX$$

$$+ \int_{-\sigma^{2m}}^0 dt \int_{-\sigma^{2m}}^0 d\xi \int_{B(\sigma)} \frac{\| u(x,t) - u(x,\xi) \|^2}{|t-\xi|^2} dx \} .$$

This work is the first approach to the regularity or partial regularity theory for the solutions to nonlinear parabolic systems of general type

$$\sum_{|\alpha| \le m} (-1)^{|\alpha|} D^\alpha a^\alpha(X, Du) + \frac{\partial u}{\partial t} = 0 .$$

REFERENCES

1.   A. Maugeri, Partial Hölder regularity for the derivatives of order $(m-1)$ of solutions to $2m$ order quasilinear parabolic system with linear growth, <u>Boll. U. M. I.   Analisi Funzionale e Applicazioni</u>, (6) <u>1-C</u> (1982) 177-191.

2.   M. Marino, A. Maugeri, L^p theory and partial Hölder continuity for quasilinear parabolic systems of high order with strictly controlled growth, <u>Ann. di Mat. Pura e Appl.</u> (to appear).

Mario Marino and Antonino Maugeri
Seminario Matematico
Università
Viale A. Doria, 6
I 95125 Catania
Italy

A MAUGERI

A new method of computing solution to variational inequalities and application to the traffic equilibrium problem

Variational inequalities in R^n generalize convex programming; in fact, if $F(x)$ is a real continuously differentiable convex function defined in a non-empty, closed, convex subset K of R^n, the problem of finding x_K in K such that

$$f(x_K) = \min_{x \in K} f(x)$$

is equivalent to that of finding a point x_K in K such that

$$\text{grad } f(x_K), (x - x_K) \geq 0 \qquad \forall x \in K .$$

Variational inequalities can express equilibrium conditions when we have, for example, a cost distribution $C(F)$ that cannot be a gradient of a function. This is the case of the traffic equilibrium problem that is visualized by the following variational inequality:

(1) find $H \in K$ such that, for all $F \in K$,

$$C(H)(F - H) = \sum_{r=1}^{m} C_r(H)(F_r - H_r) \geq 0$$

where $C(F)$ is the cost distribution which we suppose continuous and strictly monotone and K denotes the set

(2) $\quad \{ F = \sum_{r=1}^{m} \psi_{ir} F_r = \rho_i \qquad i = 1, 2, \ldots, n^2, \ F_r \geq 0 \quad r = 1, 2, \ldots, m \}$

(for more details see [1]).

In (2) ψ_{ir} are constants whose values are 1 or 0 and the variables F_r that are in one equation do not appear in the others; this structure enables us to derive the values of n^2 and so we have

(3) $\quad F_i = \rho_i - \sum_{r=n^2+1}^{m} \psi_{ir} F_r \qquad i = 1, \ldots, n^2 .$

113

By using (3) we can transform the variational inequality into the following:

(4) find $\tilde{H} \in \tilde{K}$ such that $\Gamma(\tilde{H})(\tilde{F} - \tilde{H}) \geq 0$ $\tilde{F} \in \tilde{K}$, where

$$\tilde{F} = (F_{n^2+1}, \ldots, F_m)$$

$$\tilde{K} = \{F_r \geq 0 \quad n^2+1 \leq r \leq m, \quad \sum_{r=n^2+1}^{m} \psi_{ir} F_r \leq \rho_i, \quad i = 1, \ldots, n^2\}$$

$$\Gamma(H) = (\Gamma_{n^2+1}(\tilde{H}), \ldots, \Gamma_m(\tilde{H})) \quad \Gamma_r = C_r - \sum_{r=1}^{n^2} \psi_{ir} C_i.$$

The variational inequality admits a unique solution $\tilde{H} \in \tilde{K}$ and if we suppose that there exists $\tilde{H}_0$ such that

(5) $\Gamma(\tilde{H}_0) = 0$ $\tilde{H}_0 \in \tilde{K}$

then $\tilde{H} = \tilde{H}_0$; if, on the other hand, $\tilde{H}_0$ does not belong to $\tilde{K}$ or (5) does not admit solution, $\tilde{H}$ must belong to the boundary of $\tilde{K}$. Since $\tilde{K}$ is a polyhedron of R^{m-n^2}, its boundary consists of surfaces; then we shall give a theorem that allows us to decide if $\tilde{H}$ belongs to some surface of dimension $m - n^2 - (h+k)$ with $h + k < m - n^2$. We can express a surface of dimension $m - n^2 - (h+k)$, which we shall denote by $\tilde{K}^{h,k}_{m-n^2-(h+k)}$, in this way: let us set

(i) $(S^h, J^k) = ((s_1, \ldots, s_h), (j_1, \ldots, j_k)), \quad n^2+1 \leq s_q \leq m, \quad 1 \leq j_i \leq n^2$

(ii) $I = \{n^2+1, \ldots, m\} - \{s_1, \ldots, s_h\}, \quad E = \{1, \ldots, n^2\} - \{j_1, \ldots, j_k\}$

and let us choose the indexes $l_1, \ldots, l_k \in I$ such that

(6) $F_{l_i} = \rho_{j_i} - \sum_{r \in I-\{l_i\}} \psi_{j_i r} F_r.$

If we set $L = I - \{l_1, \ldots, l_k\}$, we have

$$\tilde{K}^{n,k}_{m-n^2-(h+k)} = \{F_r \geq 0, \, r \in L, \, \sum_{r \in L} \psi_{j_i r} F_r \leq \rho_{ji},$$

$$j_i \in J^k, \quad \sum_{r \in L} \psi_{i r} \rho_i \leq \rho_i \quad i \in E\}.$$

Now let us consider the variational inequality at the surface $\tilde{K}^{h,k}_{m-n^2-(h+k)}$, that is:

$$(7) \begin{cases} \text{find } \tilde{H}^{h,k} \in \tilde{K}^{h,k}_{m-n^2-(h+k)} \quad \text{such that} \\ \Gamma^{h,k}(\tilde{H}^{h,k})(\tilde{F}^{h,k} - \tilde{H}^{h,k}) \geq 0 \quad \forall \tilde{F}^{h,k} \in \tilde{K}^{h,k}_{m-n^2-(h+k)} \end{cases}$$

where $\Gamma^{h,k}$ is the restriction of Γ on the surface $\tilde{K}^{h,k}_{m-n^2-(h+k)}$.

If there exists $\tilde{H}_0^{h,k} \in \tilde{K}^{h,k}_{m-n^2-(h+k)}$ such that

$$(8) \quad \Gamma^{h,k}(\tilde{H}_0^{h,k}) = 0$$

the following result holds:

<u>Theorem 1</u>   Let us suppose that $\psi_{j_i s_q} = 1$ when $j_i \in J_p$ and $s_q \in S_{j_i}$, where J_p is a subset (even empty) of p elements of J^k and S_{j_i} is a p subset of S^h; then $\tilde{H}_0^{h,k}$ coincides with $\tilde{H}$ if and only if

$$(9) \quad \begin{aligned} &\Gamma_r(\tilde{H}_0^{h,k}) \geq 0 \quad r \in S^k - \bigcup_{j_i \in J_p} S_{j_i} \\ &\Gamma_{\ell_i}(\tilde{H}_0^{h,k}) \leq 0 \quad i = 1, \ldots, k \\ &\Gamma_{s_q}(\tilde{H}_0^{h,k}) - \Gamma_{\ell_i}(\tilde{H}_0^{h,k}) \geq 0 \quad s_q \in S_{j_i}, \; j_i \in J_p. \end{aligned}$$

Now if (9) are not satisfied we have $\tilde{H} \notin \tilde{K}^{h,k}_{m-n^2(h+k)}$; whereas if (8) does not admit solution belonging to $\tilde{K}^{h,k}_{m-n^2-(h+k)}$, $\tilde{H}^{h,k}$ must belong to $\partial \tilde{K}^{h,k}_{m-n^2-(h+k)}$, namely to a surface of dimension $m - n^2 - (h+k+1)$ for which we can repeat the same considerations; consequently, if (8) and (9) are not satisfied for all surfaces $\tilde{K}^{h,k}_{m-n^2-(h+k)}$ with $h+k \leq m-n^2$, we can say that $\tilde{H}$ is a vertex. If we denote by $\tilde{H}_i$, $i \in P$, the vertices of K, the following result holds:

Theorem 2   The equation

$$\Gamma(\widetilde{H}_i)\,\widetilde{H}_i = \min_{j\in P} \{\,\Gamma(\widetilde{H}_i)\,\widetilde{H}_j\,\} \qquad i \in P$$

admits a unique solution that coincides with $\widetilde{H}$.

REFERENCE

1.   A. Maugeri, Applications des inéquations variationnelles au problème de l'equilibre du traffic, C. R. Acad. Sc. Paris, 295 (1982), 6-12.

Antonino Maugeri
Seminario Matematico
Università
Viale A. Doria, 6
I 95125 Catania
Italy

S SALERNO
Homogenization with degenerate constraints on the gradient

We study the following problem: let $\psi : \mathbb{R}^n \to \overline{\mathbb{R}}$ be a non-negative measurable function, 1-periodic in all the variables; let $f : \mathbb{R}^n_x \times \mathbb{R}^n_w \to \mathbb{R}$ be a non-negative function, 1-periodic and measurable with respect to x, strictly convex with respect to w. Suppose that $\{u_h\}_{h \in \mathbb{N}}$ is the sequence of the solutions of the problems

$$\underset{\substack{u \in \mathcal{F}(\Omega) \\ u = 0 \text{ on } \partial\Omega \\ |Du(x)| \leq \psi(hx) \text{ a.e. in } \Omega}}{\text{Min}} \{ \int_\Omega f(hx, Du(x)) \, dx + \int_\Omega \tilde{f} \cdot u \, dx \} \tag{1}$$

where $\tilde{f} \in \mathcal{G}$, $\mathcal{F}$ and $\mathcal{G}$ are suitable functional spaces, depending on f and $\Omega \in \text{Ap}_n$ (the collection of the bounded Lipschitzian open sets of $\mathbb{R}^n$).

Then, according to a conjecture of Bensoussan, Lions and Papanicolaou [1], the sequence u_h is expected to converge (in a suitable topology) to a function u_∞ which is the solution of a problem of the same type as (1), but with an integrand independent of x (homogenized problem). Moreover, the minima (1) would converge to the minimum of the limited problem.

The problem with non-degenerate constraint, that is, $0 < m \leq \psi \leq M < +\infty$, has been settled by Carbone [2] for ψ constant and by Carbone and the author [3] in the general case.

Under suitable assumptions, the result still holds if ψ is allowed to be zero in some region (not disconnecting the unit cube), but it remains bounded, as is proved by Carbone and the author in [4]. In all these cases, since ψ is bounded, one takes $\mathcal{F} = \text{Lip}_{\text{loc}}(\mathbb{R}^n)$, $\mathcal{G} = L^1(\Omega)$ and $u_h \to u_\infty$ uniformly.

Finally, in [5] the same authors attack the problem with ψ unbounded. Assuming

(i) $\exists \theta \in [0, \frac{1}{2})$ such that $\psi(x) = +\infty$ $\forall x \in [0, 1]^n \setminus [\frac{1}{2} - \theta, \frac{1}{2} + \theta]^n$

117

(ii) $c_1 |w|^2 \le f(x, w) \le c_2 |w|^2$ $x \in [0, 1]^n$, $w \in \mathbb{R}^n$

and taking $\mathcal{F} = H^{1,2}(\Omega)$, $\mathcal{G} = L^2(\Omega)$, they prove the following:

<u>Theorem</u> The sequence u_h of the solutions of (1) converges to u_∞ in $L^2(\Omega)$ where u_∞ is the solution of the problem

$$\underset{u \in H_0^{1,2}(\Omega)}{\text{Min}} \{ \int_\Omega w(Du(x))\, dx + \int_\Omega \tilde{f} \cdot u\, dx \} \qquad (2)$$

with

$$w(\xi) = \underset{\substack{u \in H^{1,2}([0,1]^n) \\ u - \xi \cdot x \; 1\text{-per.} \\ |Du| \le \psi \; \text{a.e.}}}{\text{Min}} \int_{[0,1]^n} f(x, Du(x))\, dx$$

Moreover, the minima (1) converge to the minimum (2).

It is worth pointing out that in general we do not have the uniform convergence of u_h to u_∞. In [5] a counterexample is given, as well as an explicit description and some important qualitative properties of the limit problem.

The main tool for the proof of the results stated above is the theory of Γ-convergence, for which we refer to De Giorgi [7]. For the application to variational functionals, see also Dal Maso and Modica [6].

REFERENCES

1.   A. Bensoussan, J. L. Lions and C. Papanicolaou. <u>Asymptotic Analysis for Periodic Structures</u>, North-Holland, Amsterdam (1978).

2.   L. Carbone. Sur un problème d'homogénéisation avec des contraintes sur le gradient, <u>J. Math. Pures Appl.</u> <u>58</u> (1979) 275-297.

3.   L. Carbone and S. Salerno. On a problem of homogenization with quickly oscillating constraints on the gradient, <u>J. Math. Anal. and Appl.</u>, <u>90</u> (1982) 219-250.

4.   L. Carbone and S. Salerno. Further results on a problem of homogenization with constraints on the gradient (to appear).

5.   L. Carbone and S. Salerno. Homogenization with unbounded constraints on the gradient, <u>Nonlinear Anal.</u> (to appear).

6.   G. Dal Maso and L. Modica.   On a general theory of variational
        functionals, Topics in Funct. Anal. (1980/81), Sc. Norm. Sup. Pisa.

7.   E. De Giorgi.   Convergence problems for functional and operators,
        Proc. Int. Meeting on Recent Methods in Nonlinear Analysis, Pitagora,
        ed. (1979).

Saverio Salerno
Istituto di Matematica
Università
I 84100 Salerno
Italy

A VALLI
Global existence theorems for compressible viscous fluids

We want to present some results concerning the existence of global solutions to the Navier–Stokes equations for barotropic compressible fluids. In particular, we have obtained a theorem which shows the existence of almost–periodic, periodic and stationary solutions (see [3], [1]) . By quoting from Serrin [2], p. 237, the problem can be written in this form:

$$(1.1) \quad \begin{cases} \rho[\, \partial_t v + (v - \nabla)\, v - b] = -\nabla[\, \bar{p}(\rho)\,] + \mu \Delta v + (\zeta + \frac{1}{3}\mu)\, \nabla \operatorname{div} v \text{ in}]t_0, \infty[\, x, \Omega, t_0 \in \mathbb{R}, \\ \partial_t \rho + \operatorname{div}(\rho v) = 0 \qquad\qquad\qquad\qquad \text{in}]t_0, \infty[\, x\Omega, \\ v\big|_{\partial\Omega} = 0 \qquad\qquad\qquad\qquad\qquad\quad \text{on}]t_0, \infty[\, x\partial\Omega, \end{cases}$$

where v and ρ are the velocity and the density of the fluid; $\bar{p}$ is the (thermodynamic) pressure, which is a known (smooth) function of ρ; b is the external force field; the constants $\mu > 0$ and $\zeta \geq 0$ are the viscosity coefficients; Ω is a bounded connected open subset of $\mathbb{R}^3$, with boundary $\partial\Omega$ smooth enough.

By $(1.1)_2$, $(1.1)_3$ the total mass of the fluid is a physical constant of the problem, i.e. we must require that

$$(1.2) \quad \int_\Omega \rho(t, x)\, dx = \bar{\rho}\, \mathrm{vol}(\Omega), \quad \bar{\rho} > 0.$$

If, in addition to (1.1) , (1.2) , we assign the initial data

$$(1.3) \quad v\big|_{t=t_0} = v_0, \quad \rho\big|_{t=t_0} = \rho_0, \quad t_0 \in \mathbb{R},$$

we can prove the following theorem:

<u>Theorem 1</u>   (Global existence)   Assume that $b \in L^2_{loc}(\mathbb{R}^+_{t_0}; H^1(\Omega))$, $\partial_t b \in L^2_{loc}(\mathbb{R}^+_{t_0}; H^{-1}(\Omega))$, $v_0 \in H^2(\Omega) \cap H^1_0(\Omega)$, $\rho_0 \in H^2(\Omega)$, $\int_\Omega \rho_0(x)\, dx = \bar{\rho}\, \mathrm{vol}(\Omega)$, $\inf\limits_{\bar{\Omega}} \rho_0(x) > 0$ and assume that

$$\|v_0\|_2^2 + \|\rho_0 - \bar\rho\|_2^2, \qquad \sup_{t \in \mathbb{R}_{t_0}^+} \int_t^{t+1} ((b)^2 \, d\tau$$

are small enough. Then there exists a unique pair (v, ρ),

$v \in L_{loc}^2(\mathbb{R}_{t_0}^+; H^3(\Omega)) \cap C_B^0(\mathbb{R}_{t_0}^+; H^2(\Omega))$ with $\partial_t v \in L_{loc}^2(\mathbb{R}_{t_0}^+; H^1(\Omega)) \cap$

$C_B^0(\mathbb{R}_{t_0}^+; L^2(\Omega))$, $\rho \in C_B^0(\mathbb{R}_{t_0}^+; H^2(\Omega))$ with $\partial_t\rho \in C_B^0(\mathbb{R}_{t_0}^+; H^1(\Omega))$,

$\rho(t, x) \geq \bar\rho/2$ in Q_{t_0}, such that (v, ρ) is the solution of $(1.1)-(1.3)$ in Q_{t_0}.

Moreover, we have that $\displaystyle\sup_{t \in \mathbb{R}_{t_0}^+} (\|v(t)\|_2^2 + \|\rho(t) - \bar\rho\|_2^2)$ is small.

Here $\mathbb{R}_{t_0}^+ \equiv]t_0, \infty[$, $Q_{t_0} \equiv \mathbb{R}_{t_0}^+ \times \Omega$, $H^k(\Omega)$ is the usual Sobolev space

endowed with the norm $\|\cdot\|_k$, $k+1 \in \mathbb{N}$, and $((b))^2 \equiv \|b(\tau)\|_1^2 + \|\partial_t b(\tau)\|_{-1}^2$.

<u>Theorem 2</u>   (Asymptotic stability)   Let (v_1, ρ_1) and (v_2, ρ_2) be two

solutions of $(1.1)-(1.3)$ in Q_{t_0} such that

$$\|v_i(t)\|_2^2 + \|\rho_i(t) - \bar\rho\|_2^2 \leq B_1, \qquad t \in \mathbb{R}_{t_0}^+, \quad i = 1, 2,$$

$$\int_{t_0}^t \|v_1\|_3^2 \, d\tau \leq B_2 + B_3(t - t_0), \qquad t \in \mathbb{R}_{t_0}^+,$$

where B_1 and B_3 are small enough. Then the difference (w, η) between

(v_1, ρ_1) and (v_2, ρ_2) satisfies

$$\|w(t)\|_0^2 + \|\eta(t)\|_0^2 \leq c(\|w(t_0)\|_0^2 + \|\eta(t_0)\|_0^2) e^{-\lambda(t-t_0)}, \quad \lambda > 0, \; t \in \mathbb{R}_{t_0}^+.$$

By means of these two theorems we are in a position to prove the following:

<u>Theorem 3</u>   (Almost-periodic, periodic and stationary solutions)   Assume that

$$\sup_{t \in \mathbb{R}} \int_t^{t+1} ((b))^2 \, d\tau$$

is small enough. If b is almost-periodic in $L^2(\Omega)$ (b is T-periodic, b is

independent of t), then there exists a solution (v, ρ) of (1.1), (1.2) in

$\mathbb{R} \times \Omega$ which has the same property.

More precise results about almost-periodicity can be found in $[1]$.

REFERENCES

1. Marcati, P. , Valli, A. Almost-periodic solutions to the Navier-Stokes equations for compressible fluids (to appear).

2. Serrin, J. Mathematical principles of classical fluid mechanics, in Handbuch der Physik, Bd. VIII/1, Springer-Verlag, Berlin Göttingen Heidelberg (1959).

3. Valli, A. Periodic and stationary solutions for compressible Navier-Stokes equations via a stability method, Ann. Scuola Norm. Sup. Pisa (to appear).

Alberto Valli
Dipartimento di Matematica
Università
I 38050 Povo (Trento)
Italy

T ZOLEZZI
Generalized dynamic programming

We consider the following optimal control problem $P(t, y)$. Minimize

$$\int_t^b f(s, x(s), u(s)) \, ds + h[b, x(b)]$$

subject to the state equation

$$\dot{x}(s) = g(s, x(s), u(s)), \quad \text{a. e.} \quad s \in (t, b), \quad x(t) = y,$$

and the constraints

$$u(s) \in U, \quad \text{a. e.} \quad s \in (t, b), \quad (b, x(b)) \in M. \tag{1}$$

Here $u \in R^m$ is the control and $x \in R^m$ the state variable.

Given some fixed real a, we consider $P(t, y)$ for $t \geq a$. We assume that b is the first time such that (1) holds, U compact and M a closed set.

The classical dynamic programming approach (see [1]) considers the value function $v(t, y)$ of $P(t, y)$ and shows that under stringent _a priori_ smoothness assumptions about v the following Bellman's equation is related to either necessary or sufficient conditions of optimality for $P(t, y)$:

$$\begin{cases} v_t + \min \{v_y' g(t, y, u) + f(t, y, u) : u \in U\} = 0 \\ v = h \quad \text{on} \quad M. \end{cases} \tag{2}$$

The restrictive assumptions we need in order to use (2) suggest that we try to extend the dynamic programming approach to a broader class of optimal control problems by considering generalized solutions of (2) obtained with the help of the generalized differential calculus of Clarke [2].

Let us assume that v is a locally Lipschitz continuous function. Explicit sufficient conditions about the data of $P(t, y)$ are known for this property. Then, if f and g are continuous, for every (t, y) and $w \in \partial v(t, y)$ the Clarke's generalized gradient of v at (t, y),

$$w_t + w_y' g(t, y, u) + f(t, y, u) \geq 0, \quad u \in U. \tag{3}$$

Moreover, if $(u^\star, x^\star)$ is an optimal pair for $P(t, y)$, then for a.e. $s \in [t, b]$ there exists $w \in \partial v(s, x^\star(s))$ such that

$$w_t + \min\{ w_x' g(s, x^\star(s), u) + f(s, x^\star(s), u) : u \in U\} = 0$$
$$= w_t + w_x' g(s, x^\star(s), u^\star(s)) + f(s, x^\star(s), u^\star(s)). \tag{4}$$

In (3) and (4) we denote by w_t the first component and by $w_x \in R^n$ the vector of the remaining components of w.

Conversely, condition (4) is sufficient for optimality of the admissible pair $(u^\star, x^\star)$ (continuity of f, g is not required).

Detailed results and relation to the generalized form of Pontryagin's maximum principle will appear elsewhere.

Related results are given in [2] section 3.7, [4] and [5]. For specific result about the linear time-optimal problem, see [3].

REFERENCES

1.   W. H. Fleming, R. W. Rishel. Deterministic and Stochastic Optimal Control, Springer (1975).

2.   F. H. Clarke. Optimization and Nonsmooth Analysis, Wiley (1983).

3.   F. Mignanego, G. Pieri. On a generalized Bellman's equation for the optimal-time problem, System and Control Letters (to appear).

4.   V. Barbu, G. Da Prato. Hamilton-Jacobi Equations in Hilbert Spaces, Pitman (1983).

5.   P. L. Lions. Generalized Solutions of Hamilton-Jacobi Equations, Pitman (1982).

Tullio Zolezzi
Istituto Matematico, Università
Via L. B. Alberti, 4
I 16132 Genova
Italy